Digital Skill Development for Industry 4.0

The Fourth Industrial Revolution, Industry 4.0, is changing the world, and digital transformation technology tools have impacted every walk of life. The nature of work and careers is changing fast – and in the future, the right skills will be prized over academic qualifications. Students must develop various skills, especially technology skills, to become the workforce of the future; the onus of developing these skills falls on educational institutions. The development of innovation and ideation skills in students is a must for them to productively contribute to a future economy. They should have the capability to translate ideas into solutions, products, and systems that are scalable, practically feasible, and adaptable.

To prepare students for this digital world, *Digital Skill Development for Industry 4.0* contains chapters on developing:

- Professional skills, cognitive learning
- Critical thinking skills among modern learners
- Motivation and student engagement
- Invoking interest of students
- Student engagement in e-learning
- Contextual learning
- Experimental learning
- Multidisciplinary learning

The book is written for the academic and administrative community of universities and colleges who intend to equip their modern learners with the skills of Industry 4.0. It is also useful as a reference for undergraduate and postgraduate courses, as well as doctoral programs. The book can also serve as a comprehensive reference guide for researchers in the development of tools incorporating new skills and technologies. Practitioners who are interested in introducing digital skill development for Industry 4.0 and its tools to solve real-world problems can also use this book.

Digital Skill Development for Industry 4.0

Edited by

P. Kaliraj, G. Singaravelu, and T. Devi

CRC Press
Taylor & Francis Group
AN AUERBACH BOOK

First edition published 2025
by CRC Press
2385 NW Executive Center Drive, Suite 320, Boca Raton FL 33431

and by CRC Press
4 Park Square, Milton Park, Abingdon, Oxon, OX14 4RN

CRC Press is an imprint of Taylor & Francis Group, LLC

ISBN: 978-1-032-53197-7 (hbk)
ISBN: 978-1-032-82509-0 (pbk)
ISBN: 978-1-003-50489-4 (ebk)

DOI: 10.1201/9781003504894

Typeset in Garamond
by codeMantra

Prof. P. Kaliraj *dedicates this book to*
esteemed Bharathiar University,
his father Mr. M. Perumal,
mother Mrs. P. Rathinammal,
wife Dr. Vanaja Kaliraj, and
grandchildren Prithika Karthikeyan,
Kayan Karthikeyan,
Ayaan Prashanth, and
Akira Prashanth.

Prof. G. Singaravelu *dedicates this book to*
Department of Education, Bharathiar University,
his father Mr. R. Govindarajan,
mother Mrs. G. Sankarambal,
wife Mrs. G. Jeevakumari,
son-in-law Dr. P. Ramachandran,
daughter Dr. S. Yazhini,
son Er. S. Sabarirajan, and
grandson R. Jaswickram.

Prof. T. Devi *dedicates this book to*
Department of Computer Applications, Bharathiar University,
her mother Mrs. Suseela Thirupathi,
mother-in-law Mrs. Singari Doraiswamy,
husband Mr. Ravi Doraiswamy,
son-in-law Mr. Vishnu Prakhash Devaraj
daughter Mrs. Deepiga Vishnu Prakhash,
son Mr. Surya Ravi, and
grandson V. Deera.

Contents

Acknowledgments

From Dr. P. Kaliraj
First and foremost, I express my sincere gratitude to Hon'ble Shri Banwarilal Purohit, former governor of Tamil Nadu, India, who was instrumental in organizing the conference on Innovating Education in the era of Industry 4.0 on 19–20 December 2019 in Ooty, which paved the way for further work in the Industry 4.0 knowledge world, and Hon'ble Shri. R.N. Ravi, governor of Tamil Nadu, India, who was instrumental in organizing the Conference of Vice-Chancellors of State, Central, and Private Universities in Tamil Nadu at Raj Bhavan, Ooty during 25–26 April 2022, which paved the way to redefine the future course of action particularly in education and sustainable developments.

My heartfelt thanks go to the Hon'ble Chief Minister of Tamil Nadu, India, and the Hon'ble Minister for Higher Education, Government of Tamil Nadu. I thank the Principal Secretary to Government, Higher Education Department, Government of Tamil Nadu.

I would like to express my thanks to Secretary to Governor, and Deputy Secretary to Governor, Universities Governor's Secretariat, Raj Bhavan, Chennai.

I thank my wife Dr. Vanaja Kaliraj, my son-in-law Mr. M. Karthikeyan, daughter Mrs. Ratna Priya Karthikeyan, son Dr. K. Siva Prashanth, daughter-in-law Dr. Ratna Priya Prashanth, grand-children Prithika Karthikeyan, Kayan Karthikeyan, Ayaan Prashanth, Akira Prashanth, and family members for their support and being patient.

From Prof. G. Singaravelu
I submit my sincere gratitude to Prof. P. Kaliraj, Hon'ble Former Vice-Chancellor of Bharathiar University, who endeavored to solve the problems by using matters and materials of the book named Digital Skill Development for an Industry 4.0 World which paved the way to provide knowledge on skills needed to face the Volatile, Uncertain, Complex and Ambiguous (VUCA) world. His herculean task and tireless motivation, continuous acceleration and encouragement, created the book.

I convey my sincere thanks to the Vice-Chancellor and Registrar for administrative support. I shower my gratitude to the authors of the chapters for their contribution of chapters, continuous cooperation in improvising the chapters as and when requested, and for timely communication. I have to thank wholeheartedly toward all the experts who served as reviewers for making quality and swift reviews.

We wish to thank Mr. John Wyzalek, Senior Acquisitions Editor, Taylor & Francis/CRC Press, who believed in the idea of this book and helped us in realizing our dream.

I express my sincere and Indispensable thanks to Prof. Dr. T. Devi, Dean-Research, PSG Institute of Management, PSG College of Technology, former Professor and Head, Department of Computer Applications, School of Computer Science and Engineering, Bharathiar University.

Thanks are due to Dr. A. Sivakumar, Assistant Professor (Guest), Department of Education, for his continuous technical effort and academic support in tailoring the book.

From Prof. T. Devi

I record my sincere thanks to Prof. P. Kaliraj, Hon'ble Former Vice-Chancellor of Bharathiar University, who identified the gap in the knowledge world when the professor searched for a book on Industry 4.0 and triggered the process of writing and editing books in the Industry 4.0 series. His continuous motivation during the lockdown period due to COVID-19, sensitization, and encouragement is unmatchable.

I express my profound thanks to the Vice-Chancellor and Registrar for the administrative support. Heartfelt thanks are due to the authors of the chapters for their contributions, continuous cooperation in improvising the chapters as and when requested, and for timely communication. I thank all the expert members who served as reviewers for providing a quality and swift review.

We wish to thank Mr. John Wyzalek, Senior Acquisitions Editor, Taylor & Francis/CRC Press, who believed in the idea of this book and helped us in realizing our dream.

Special thanks are due to Mr. Glenon Butler, Production Editor, Taylor & Francis/CRC Press, Florida, for his earnest co-ordination and Ms. Kathrin Immanuel, Project Manager, codeMantra for her stupendous support.

Thanks to Ms. Stephanie Kiefer, Editorial Assistant, Taylor & Francis Group/CRC Press, Boca Raton, FL, for her efforts and communications during the project.

Thanks are due to Mrs. K. Kowsalya, Assistant Registrar; Mr. R. Karthick, Assistant Section Officer; Mr. A. Prasanth and Mr. Vijayakumar of the office of the vice-chancellor and staff of the office of the Registrar of Bharathiar University, India.

Special thanks to Dr. A. Sivakumar, Assistant Professor (Guest), Department of Education, who supported the plagiarism check and generated excel sheets and follow-up work.

I thank my husband Mr. D. Ravi, son-in-law Mr. D. Vishnu Prakhash, daughter Mrs. R. Deepiga, son Mr. R. Surya, grandson V. Deera, and family members for their encouragement and support.

Preface

The Fourth Industrial Revolution, Industry 4.0 is changing the world around us, and digital transformation technology tools have impacted every walk of life. The nature of work and careers is changing fast – and in the future, the right skills will be prized over academic qualifications. Industries need life-long learners with ever-relevant skills. Students must develop various skills, especially technology skills, to become the workforce of the future; the onus of developing these skills falls on educational institutions. The educational institutions are trying to prepare work-ready workforce, reduce on-the-job training cost, and make learners more employable for Industry 4.0. Global citizenship skills, innovation and creativity skills, technology skills and interpersonal skills are emphasized for learning in this era by the World Economic Forum. This book details the skills needed to face the Volatile, Uncertain, Complex, and Ambiguous (VUCA) world. Moreover, it discourses on professional skills, cognitive learning, critical thinking skills among modern learners, motivation and student engagement, invoking interest of students, student engagement in e-learning, contextual learning, experimental learning, multidisciplinary learning, team collaboration using digital technologies, digital evaluation tools, student assessment, teacher capacity building, and gender-responsive digital skills training.

The World Economic Forum has emphasized four categories of skills for learning in this era. They are global citizenship skills, innovation and creatively skills, technology skills, and interpersonal skills. Creating future citizens would require education institutions to help children in developing global awareness, comprehension of global interdependencies of issues, and their role and responsibilities within the global community. Emerging technologies such as virtual classrooms, social media, and cyber-physical systems can be used for global citizenship education. The development of innovation and ideation skills in students is a must for them to productively contribute to a future economy. They should have the capability to translate ideas into solutions, products, and systems which are scalable, practically feasible, and adaptable. Skills such as curiosity, creativity, critical thinking, problem-solving, and systems analysis enable innovation. Students must develop technology skills to become the workforce of the future; the onus of developing these skills falls on educational institutions. Interpersonal skills also known as people skills – qualities like leadership, social influence, empathy, teamwork, emotional intelligence – are among the high-demand interpersonal skills of the future. These skills can help students appreciate different perspectives and viewpoints and develop long-lasting meaningful relationships with fellow beings.

The Fourth and Fifth Industrial Revolutions are affecting the roles that Indian universities and colleges prepare students for, and educational institutions are committed to help produce the workforce for this new world and the student experience to match it. Digital transformation technologies are the need of the hour. In order to impart this knowledge across arts and science UG and PG programs, Bharathiar University had designed guidelines for Curriculum 4.0 and had revamped the syllabi for all subjects, intertwining Industry 4.0 and 5.0 tools into various disciplines such as Science, Social Science, Arts, and Education.

Today, the rate at which the transformation is happening is very disruptive in nature, and also exponential changes are being witnessed. We have already witnessed how many conventional business models and technologies got disrupted in the last few years. Digital transformation technologies such as artificial intelligence and machine learning, augmented reality, virtual reality, and mixed reality are mutating very fast like virus. No one is able to predict properly the impact of digital transformation technologies on education.

With the adoption of digital technology by business firms, healthcare, start-ups, and almost all government institutions across the globe, an environment has evolved whereby no walk of life can be imagined without digital technology. Worldwide, stakeholders are working toward improving the industry academia collaboration for a more joined up approach leading to excellence in higher education, work ready workforce, and quality research outputs for the benefit of the industry and the academia. *Providing high quality skills training for young people is a priority in every country. Collaborations and expert networks are established to bridge the skills gap and create employable skills among the workforce.* Toward enhancing the skills and employability of young people in the era of Industry 4.0, academic institutions and industries are establishing collaborations and *intercountry agreements are created.* The book lime lights on *skill development* for alleviating the challenges of modern learners. Due to advancements in technology, education involves a blended, online digital learning. Teachers will turn out to be facilitators in futuristic learning.

As the educational institutions are innovating education and especially skill development of students in the era of Industry 4.0 and 5.0, this book comes at the right time to help teachers, scholars, and students. The digital transformation technological tools such as artificial intelligence, mobile computing, cloud computing, and augmented reality are changing the way teaching and learning happen in the modern era. *This book would be a boon for members of board of studies, technocrats, policy makers, pre-service teachers, in-service teachers, skill trainers, scholars, and students.*

The revenue forecast of these digital transformation technologies in teaching and learning, in other words, in EdTech in 2030 is expected to be around USD

429.5 billion, and the growth rate from 2022 to 2030 in terms of CAGR is expected to be 16.5%, thus creating a good market for the skilled professionals in this area. There is a wider gap in the knowledge resources cutting across digital transformation technologies and the education of arts, science, social science, pharma, and education disciplines. To fill this gap and to prepare the future pillars of our globe to face the volatile, uncertain, complex, and ambiguous (VUCA) world, and to help the academic community, knowledge resource on *Digital Skill Development for an Industry 4.0 World* has been initiated.

This book contains chapters on professional skills, cognitive learning, critical thinking skills among modern learners, motivation and student engagement, invoking interest of students, student engagement in e-learning, contextual learning, experimental learning, multidisciplinary learning, team collaboration using digital technologies, digital evaluation tools, student assessment, teacher capacity building, and gender-responsive digital skills training. The book would greatly help the teachers and the future pillars of the globe.

The book opens with a chapter, "Need of Professional Skills at Present Scenario" which explains the need and importance of techno-skills and techno-pedagogy as the basic professional skills of teachers at all levels in a precise manner. "Cognitive Learning" comes second. Out of the nine cognitive skills a human has, six of the skills namely, attend, listen, think, understand, solve, and respond are included in Cognitive Learning. This chapter provides a detailed view on the definition, genesis, types, theories (literature survey), elucidation, impact evaluation, limitations, applications, conclusion, and bibliography on Cognitive Learning and explores alternative techniques that may enhance Cognitive Learning ability. "Professional Skills" detail on the skills that empower an individual with ability to take on new challenges and opportunities. "Fostering Critical Thinking Skills in Modern Learners: A Peep into Higher Educational Institutions" is a chapter that explores the key factors and instructional practices that are needed in fostering critical thinking skills among modern learners in higher educational institutions and analyzes how critical thinking skills assist the modern learners in the decision-making process. It also makes a study on the challenges faced by higher educational institutions in developing critical thinking skills among modern learners.

Student engagement refers to the degree of attention, curiosity, interest, optimism, and passion that students show when they are learning or being taught, which extends to the level of motivation they have to learn and progress in their education. "Creating Interest in Students – A Gaze" discusses on how to make a student flourish in his academic life with beautiful feathers. It details on how to make individuals more responsible by taking care of the societal needs by enhancing their quality through effective learning process with the use of embedded technologies. "Role of Need Analysis in Invoking Interest of Students to Learn English" discusses on how to provide a fruitful learning experience and keep the learners self-motivated and deeply involved throughout the learning process of English as a professional language. "A Systematic Literature Review on Student Engagement in

E-Learning: Challenges and Prospects in Higher Education" examines past research findings to offer a comprehensive overview of the studies on students' engagement in online learning, identifies challenges faced by teachers and learners, and suggests how effective student engagement is possible. "Contextual Learning" describes the foundational theories of contextual learning and the strategies for its implementation in the learning process. "Experimental Learning: The Source of Development" takes the reader through the process of real time learning, conceptual apprehension, and active experimentation.

"Multidisciplinary Learning" explores how a multidisciplinary approach to education will help synergize learning into an all-inclusive unit and enable the students to see real correlations between varied subjects. It shows how an interdisciplinary approach will enhance the learning efficiency of the students and broaden their thinking horizons. "Empowering Team Collaboration Using Digital Technologies: An Innovative Teaching Style in Higher Education" shows how the teacher and learners work together as structural interdependence to stimulate creativity, innovation, and generate new ideas using different soft skills, and discusses on methods to enhance team collaboration. "Digital Evaluation Tools" take a look into the digital evaluation tools available nowadays and how they work. "Student Assessment and Evaluation in the Disruptive Model of Teaching and Learning in Indian Higher Education: A Critical Analysis" intends to provide a critical outlook to the problem of student assessment and evaluation in the disruptive model of teaching and learning in Indian higher education. Ways to balance and ameliorate the gaps and differences in implementing digital assessment and evaluation in higher education in India have also been suggested.

"Teacher Capacity Building in High-Performing Digital Education Ecosystem" provides some strategic priorities, an initial framework describing digital concepts toward teaching quality and innovating education in a new era. "Acceleration of Active Learning in Teaching Learning Transactions" illuminates expediting the teaching and learning using Active Learning (AL) in the classroom transactions. "Revitalizing Education: Building Educational Resilience in the Post-Pandemic Era" describes the need and importance of blended and hybrid learning during the post-pandemic. "Translation: An Efficient Tool in the Teaching of Foreign Languages: A Special Reference to the French Language" deliberates upon the advantages of learning a foreign language with a special reference made to the teaching learning of French language in India. "Immediate Effect and Future Perspective of COVID-19 Pandemic on School Education" highlights the various initiatives taken by the Government of India to educate the school-going children.

"Multilingualism on Global Education and Language Learning" investigates the function of multilingualism in education and language learning. It foresees the needs and importance of languages and urges the students to incorporate multiple languages into their curriculum budding the three-language formula. It also implores how globalization has altered the history of education stressing more importance on the integration of humanities, arts, science, and STEM disciplines

(Science, Technology, Engineering, and Mathematics). Access to education is no longer limited to the local geography. Chapter entitled "E-Learning – A Way for Bridging the Digital Divide In Andaman and Nicobar Islands: Opportunities and Challenges" discusses the various e-learning methods and focuses on the avenues that digital learning provides and the challenges it poses to the natives of Andaman and Nicobar Islands. "Digital Financial Transaction Practices among Urban and Rural Women" describes the digital literacy skills that involve skills on digital technologies such as internet platforms, social media, and mobile devices. This chapter details the study of the status of the gender digital divide in rural women's access to and use of computers and internet. It also addresses the principles that urge gender-responsive digital skills training.

The primary audiences for this book are students, scholars, teachers, and teacher educators of higher education institutions who need a reference resource, educationists, and policy makers who would introduce digital skill development for Industry 4.0 and tools in educational institutions. The students, scholars, and teachers can be from arts and science universities and teacher education universities. Practitioners who are interested in introducing digital skill development for Industry 4.0 and its tools to solve real-world problems can also use this book for reference. It is also meant for the administrative community of universities and colleges who intend to equip their modern learners with the skills of Industry 4.0.

Highlights of the Book

The highlights of the book include the skills needed in the disruptive digital transformation era such as professional skills, critical thinking skills, life-long learning skills, arising and increasing the interest of learning, active learning, blended learning, digital evaluation tools, teacher capacity building, and gender-responsive digital skills training.

How to Use the Book

The method and purpose of using this book depend on the role that you play in an educational institution or in an industry or depend on the focus of your interest. We propose six types of roles: student, software developer, teacher, member of board of Studies, researcher, and administrator of educational and training institutions.

If you are a student: Students can use the book to get a basic understanding of professional skills. Critical thinking skills in modern learners are dealt with in higher educational institutions environment. The fundamental concepts and the frameworks of contextual learning can be learnt from "Contextual Learning". The world is evolving at a rapid pace, and those who constantly up skill and adapt will be the ones to thrive in this dynamic environment. Skills are the bridge that connects education to employment, and equipping individuals with relevant skills is crucial for career progression. Methods to simulate creativity, innovation, and idea generation can be studied from the chapter entitled "Empowering Team Collaboration Using Digital Technologies: An Innovative Teaching Style in Higher Education". The knowledge, need, and importance of blended learning can be obtained from the chapter entitled "Revitalizing Education: Building Educational Resilience in the Post-Pandemic Era" and active learning in the classroom from "Acceleration of Active Learning in Teaching Learning Transactions". The various e-learning methods and the avenues and challenges of digital learning can be learned from "E-Learning – A Way for bridging the Digital Divide in Andaman and Nicobar Islands: Opportunities and Challenges".

If you are a software developer: Software developers can use the book to get a basic understanding of the professional skills from chapters like "Need of Professional Skills at Present Scenario and Professional Skills". Student engagement in online learning will inspire them to develop learning platforms or mobile apps that would improve student engagement in e-Learning. Various learning methods such as contextual, experimental, and cognitive can be understood from the Chapters like "Contextual Learning", "Experimental Learning: The Source of Development", and "Cognitive Learning" will aid software developers to design and develop digital technology-based learning systems to supplement educational materials which can help students in optimizing their learning experience by enhancing student engagement skills. The chapter on "Empowering Team Collaboration using Digital Technologies" would help in designing systems that would aim at providing the team collaborative skills using technologies.

If you are a teacher: The book is useful as a text for several different university level and college-level undergraduate and postgraduate courses besides doctoral programs. The chapter "Need of Professional Skills at Present Scenario" provides the need and importance of techno-skills and techno-pedagogy as the basic professional skills of teachers. The instructional practices such as online learning, blended learning, and digitized learning that would foster critical thinking skills among modern learners are given in "Fostering Critical Thinking Skills in Modern Learners: A Peep into Higher Educational Institutions" and methods to invoke interest of students can be learned from "Role of Need Analysis in Invoking Interest of Students to Learn English". The disruptive model of teaching and learning, digital evaluation tools, critical outlook on student assessment and evaluation, teacher capacity building, acceleration of active learning, blended learning, and e-learning detailed in this book will be very useful to the teaching community.

If you are a member of a board of studies: As the world moves fast with Industry 4.0 and Industry 5.0 around, creative people emerge with applications involving disruptive technologies like artificial intelligence, augmented reality, and learning analytics. Education imparted to students of this era should be with a focus on practical skills and hence a curriculum revisit might be essential. The book is useful for a board of studies member while framing the syllabus for new courses that discuss imparting skills in the era of disruptive technologies in the disciplines of arts or science or social science. The chapters in this book that discuss Cognitive Learning, professional skills, contextual learning, multidisciplinary learning, active learning, blended learning and translation, and multilingualism can be used as a foundation to frame courses in the education discipline. The chapters on e-learning, examination of digital skill practices, digital evaluation tools, critical outlook on student assessment and evaluation in the environment of disruptive technologies will be useful for designing curriculum for new courses that would aim to impart digital skills based on the new technologies.

If you are a researcher: Research is always a vital area where innovation is required and created, or produced. The research work carried out by universities and research institutions should be made available to society so that it benefits the common man. Innovations in the field of disruptive technologies like artificial intelligence and augmented reality can change the way people live. Understanding digital skills to handle financial transactions is important and an analysis on the digital skill practices among urban and rural women for handling financial transactions using qualitative method can be understood by the research scholars from the chapter entitled "Digital Financial Transaction Practices among Urban and Rural Women". The chapters "A Systematic Literature Review on Student Engagement in E-Learning", "Empowering Team Collaboration Using Digital Technologies", "Digital Evaluation Tools", "Student Assessment", and "Evaluation in the Disruptive Model of Teaching and Learning in Indian Higher Education" would help a researcher to comprehend the recent developments in digital skills and e-learning and to see where his/her research is. Hence, this book can serve as a comprehensive reference guide for researchers in the development of tools incorporating the new skills and technologies.

If you are an administrator of an educational and training institution: Administrators of educational programs are constantly on the strive to combine human and material resources to supervise, plan, strategize, and implement structures to execute an education system. This book will help them with components for designing degree programs, courses, curricula, and syllabi incorporating the digital skills for using internet, social media, and mobile devices for education, e-learning, various methods of e-learning, avenues that digital learning provides. To develop faculty development programs in high-performing digital education ecosystem, this book will be very useful. The chapters "Team Collaboration Using Digital Technologies", "Digital Evaluation", and "Student Assessment and Evaluation in the Disruptive Model of Teaching and Learning in Indian Higher Education: A Critical Analysis" discuss the necessary skills in the disruptive technology era and would help educators to design their courses with the latest trends in these areas.

Editors

Prof. P. Kaliraj, Hon'ble former vice-chancellor, Bharathiar University, a visionary and an eminent leader leading big academic teams, has had more than three decades of teaching and research experience. He has held various renowned positions, such as officiating vice-chancellor of Anna University, head of the Centre for Biotechnology of Anna University, dean of faculty at A C College of Technology, and member of the syndicate for two decades at Anna University. Prof. Kaliraj had research collaborations with the National Institute of Health in Maryland, USA; Glasgow University in Scotland, UK; and the University of Illinois in Rockford, USA. University Grants Commission BSR Faculty Award and the Lifetime Achievement Award from the Biotechnology Research Society of India adorned the professor. Forty-two scholars were gifted to receive the highest academic degree under his distinguished guidance. His remarkable patent in the area of filariasis is a boon in healthcare and saving the lives of mankind. He is a great motivator and very good at sensitizing the faculty, scholars, and students toward achieving academic excellence and institutional global ranking. Prof. Kaliraj is a recipient of the Life Time Achievement Award and Sir J.C. Bose Memorial Award for his outstanding contribution to higher education – research. Professor promoted Industry 4.0 in Bharathiar University

Dr. G. Singaravelu is a former professor and head of the Department of Education, director in charge of UGC-HRDC, and senate member of Bharathiar University, Coimbatore, South India. He served as a member – syndicate, dean of Education, and Finance Committee member of the Bharathiar University. He has 35 years of teaching experience and he is only the professor awarded DLitt, prestigious highest degree in the university within southern states of India. He also served as coordinator for BEd programs and CRPF in the same university. He is an editorial committee member in the manager's *Journal on School Educational Technology, Journal of English Language Teaching*, and Research Demagogue. The NCERT crowned him its meritorious national award recognizing his innovative gadget in the field of Mathematics. He has been honored with Abdul Kalam Award as a best teacher and Asiriyar Chemmal Award has been conferred to him by the Government of Tamil Nadu. He has so far published 25 books besides more than 195 research papers. He has so far guided six PhD scholars in Educational Technology and 28 PhD scholars in Education. He participated and presented papers in conferences held at the University of Pennsylvania, Philadelphia, USA, University of Cologne, Germany, Brock University, Canada, University of Malaya, Malaysia, Infrastructure University, Kuala Lumpur, Nanyang University, Singapore, Dubai, and Sharjah. He is a visiting team member of NCTE, NAAC, selection committee member of principals and faculty members and inspection committee member of TNTEU. He has been acting as a member of INFITT, ISFTE, Research Advisory Board in UGC-HRDC, Bharathidasan University, Trichy, and Board of Studies at various universities.

Prof. T. Devi, PhD (UK), Dean-Research, PSG Institute of Management, PSG College of Technology, former professor and head of the Department of Computer Applications, former Director of the Centre for Research and Evaluation, former dean of Research of the Bharathiar University, focuses on state-of-the-art technology that industries adopt in order to make students ready for the future world. She is a Gold Medalist (1981–1984) from the University of Madras and a Commonwealth Scholar (1994–1998) for her PhD from the University of Warwick, UK. She has more than three and a half decades of teaching and research experience from Bharathiar University, Indian Institute of Foreign Trade, New Delhi, and University of Warwick, UK. Prof. Devi is good at team building and setting goals and achieving them. Her research interests include integrated data modeling and framework, meta-modeling, computer-assisted concurrent engineering, and speech processing. Prof. Devi had visited the United Kingdom, Tanzania, and Singapore for academic collaborations. She had received various awards including Commonwealth Scholarship and Best Alumni Award from PSGR Krishnammal College for Women (PSGRKCW), Proficiency Award from PSG College of Technology and awards from Bharathiar University for unblemished services, serving for BU-NIRF, Curriculum 4.0, and Roadmap 2030, and guided 23 PhD scholars.

Contributors

Vu Thi Lan Anh
Hanoi Law University
Hanoi, Vietnam

S. Arulsamy
Department of Education
Bharathiar University
Coimbatore, Tamil Nadu, India

A. Benjamin
Department of Education
Bharathiar University
Coimbatore, Tamil Nadu, India

A. Edward William Benjamin
Department of Educational Technology
Bharathidasan University
Tiruchirappali, Tamil Nadu, India

Ton Quang Cuong
Faculty of Education Technology
University of Education
Vietnam National University
Hanoi, Vietnam

K. Kamala Devi
Sri Sarada College of Education
Salem, Tamil Nadu, India

Antara Dey
Department of Education
Tezpur University
Assam, India

Jijish Elias
Department of Educational Technology
Bharathidasan University
Tiruchirappali, Tamil Nadu, India

Cynthia George
Department of French
Providence College for Women
Nilgiris, Tamil Nadu, India

Dinh Thi Phuong Hoa
Faculty of Legal Foreign Languages
Hanoi Law University
Hanoi, Vietnam

Kamalaveni
Department of Women Studies
Bharathiar University
Coimbatore, Tamil Nadu, India

A.R. Saravana Kumar
Department of Education (DDE)
Alagappa University
Karaikudi, Tamil Nadu, India

Hima Bindu Maringanti
Department of Computer Application
Maharaja Siriam Chandra Bhanja University
Odisha, India

M. Mirunalini
Department of Educational Technology
Bharathidasan University
Tiruchirappalli, Tamil Nadu, India

N. Mohanapriya
Department of Mathematics
Kongunadu Arts and Science College
Coimbatore, Tamil Nadu, India

Raj Kumar Pal
Department of Education
Tezpur University
Assam, India

R. Pavitra
Department of Women Studies
Bharathiar University
Coimbatore, Tamil Nadu, India

P. Janardhana Kumar Reddy
Department of Education
Bharathiar University
Coimbatore, Tamil Nadu, India

K. Revathy
Department of Education
Bharathiar University
Coimbatore, Tamil Nadu, India

Nil Ratan Roy
Department of Education
Tezpur University
Assam, India

Meenakhi Sahu
Department of Computer Application
MSCB University
Baripada, Odisha, India

Happy Saikia
Department of Education
Tezpur University
Assam, India

Parimal Sarkar
School of Education
Netaji Subhas Open University
Kalyani, Nadia, West Bengal, India

Tayum Saroh
Department of Education
Rajiv Gandhi University
Doimukh, Arunachal Pradesh, India

G. Singaravelu
Department of Education
Bharathiar University
Coimbatore, Tamil Nadu, India

A. Sivakumar
Department of Education
Bharathiar University
Coimbatore, Tamil Nadu, India

Priya Topno
Department of Education
Salesian College Siliguri
Siliguri, West Bengal, India

R. Sengamalam @ Vaanathi
Department of Education
Directorate of Distance Education
Madurai Kamaraj University
Madurai, Tamil Nadu, India

V. Vanitha
Department of Education
Bharathiar University
Coimbatore, Tamil Nadu, India

K. Venkatesan
Indira Gandhi National Tribal University
Anuppur, Madhya Pradesh, India

L. Vidyarani
Department of Mathematics
Kongunadu Arts and Science College
Coimbatore, Tamil Nadu, India

Nguyen Thi Linh Yen
University of Languages and International Studies
Vietnam National University
Hanoi, Vietnam

Chapter 1

Need of Professional Skills in the Present Scenario

R. Sengamalam @ Vaanathi

Introduction

A Professional is a person who is competent by training and practices in his specialized field during study for his future service without any compromise in his profession ever. A Profession requires one to be well prepared for the selected field, dedicated to service, trained in skills, and appropriate practices during study under an experienced and qualified mentor, and focused on a job with a full disciplinary culture without deviating from the ethical and moral path. A profession is otherwise defined as the study of particular knowledge and skills that are necessary to perform that role within the profession and it holds with strict codes of conduct. A profession not only requires possessing specified skills, knowledge, training, and durational practices but also needs possession of a determined goal of achievement in the selected career and for the organizational development along with a professional commitment and involvement. Professional skills are important for successful professional networking of any profession for one's career development and to make the field competent by interacting, sharing, transmitting, receiving, and responding to views with others for his own and others' development and effective implementation for further development. Usually, a skill is the capability to do any task with some predetermined results within a limited period, by using appropriate energy and required resources productively. Similarly, professional skills are the capability to do the task concerning their profession with predetermined output over time by using excellent professional skills inappropriate with content or subject and using effective resources for better results. In simple, professional skills are

DOI: 10.1201/9781003504894-1

capabilities that can help a person to succeed in his field and they describe a habit, personality trait, or ability that positively affects one's performance in the workplace and his working style with confident expertise.

Objectives

In this present digital era, it is important to have and practice different and innovative teaching strategies for today's learners. It is a well-known fact that 'no technology can replace an effective teacher' and hence the teacher does not have a negligible role when practicing techno-pedagogical skills. A teacher is one of the professionals and teaching is the king of all professions, it is required to have sound and effective professional skills in teaching. The integration of all available technologies in classroom teaching was initiated before the pandemic situation itself, but it did not compel all the teachers to take part in their role like this situation. Of course, even though teachers are technologically sound, they may not be compelled to execute their classroom teaching by incorporating all those techno-inputs into their students. This chapter attempted to address the need for professional skills among the teachers in this present scenario to some extent.

Professional Skills

As per Alison Doyle (2024), the essential professional skills are listed broadly as

- **Communication**: This is one of the important professional skills including written, verbal, and non-verbal communication. Other communication skills hold with:
  - advocating
  - asking for help or advice
  - brainstorming
  - building buy-in to an idea
  - business writing
  - dealing with difficult people
  - facilitating
  - handling working place politics
  - handshaking
  - interviewing
  - using information and communication technologies
  - managing the positive relationship with the others
  - listening
  - networking
  - persuasion

  - precise talk
  - social skills
  - using simple vocabularies
  - clear and neat write-up
- **Public Speaking:** It is required very good public speaking for all professions; professionals are required to be capable of speaking to others whether they belong to their field or not, clearly and present their information to others effectively. Hence, the following are skills important to those who have to present a professional's view in public:
  - articulation
  - confidence
  - creating the creative slides for his presentation
  - good posture
  - appropriate projection
  - accepting criticism and giving feedback
  - social skills
- **Teamwork:** A profession is not a single-man army and it is needed to work in some sort of a group to achieve a goal by collaborative ideas and involvement. Hence, as a professional, he must also possess the interpersonal skills to get along with others to complete any task of his career since it is required to share responsibility with others by communicating with others effectively and to be an achiever of a common goal in his field. Hence, it is required to have the following skills as an effective team member:
  - good conflict manager
  - tendency to resolve any conflict
  - move on negotiation
  - should build relationship
  - approach to building team spirit
  - have the capacity to form a team
- **Time Management**: Any profession is not a single-task job and it holds with multi-tasking actions and activities to arrive at the common goal of its field. Hence, he is required to manage to time by employing drawing organizational skills to plan his timing so that he can complete each task effectively. To be a good time manager, it is required to have the following skills:
  - Attention to all details
  - Have intrinsic motivation
  - Plan to have deadlines
  - Possess project management
  - Move on with punctuality
  - Tend to self-starting
- **Leadership:** Apart from the role of an individual in any profession, leadership skills are so important whether he is working as a team member in a team or manager or leading a team. It is required to lead the task of his profession

and it should be realized that leadership is not a position; it is a responsibility and commitment to his profession. Hence, a few of the leadership skills are listed as

- Being Accountable
- Tendency to budget
- Be calm under any unexpected situation or pressure
- Ready to coach others if required
- Coordinate the available resources
- Make good and speedy decisions if the situation arises
- Be a goal-setter
- Have developmental mind-setting
- Gather productive information required for further action
- Influence others for directed task
- Manage the time and resources available for better output
- Be a mentor for others
- Plan your work by the availability of different resources
- Behave politely
- Move on with positivity
- Do the work by prioritizing the importance of them

- **Flexibility**: All professions have their code of conduct and procedures to construct their action for their respective field. Simultaneously, there is a need for flexibility and the ability to change or alter the action based on the requirements and to derive some expected result before or after the scheduled duration. Hence, it is necessary to understand the different perspectives and to adjust the work culture according to the needs of the time for their profession. Some of the flexibilities that are required in any professional activities are:
  - Able to change the mindset
  - Adapting the conditions and availabilities
  - Analyzing the condition for further action
  - Should have anger management
  - Possess patience
  - Have good perspectives
  - Capacity to solve problems
- **Personal Skills**: Personal Skills are also pronounced as interpersonal skills which are also called soft skills which enable professionals to work well and effectively expel their potentialities. It is a well-known truth that professional skills have to be realized by enriched personal skills. Hence, the following are some of the personal skills that are to be developed, retained, and updated as per the nature of work and need of the field accordingly
  - to be a good career planner and its management
  - have positive attitude
  - free from emotional agitations and anxiety
  - willingness to accept the faults and criticism

- willingness to learn and update the knowledge
- to be a creative personality
- be a good critical thinker
- develop a tendency to adapt to the situations and resources available
- possess emotional intelligence
- have boundaries both in personal and professional
- have professional ethics
- practice honesty, integrity, and transparency in all the actions
- move on with perseverance
- believe in practicability
- respect all
- have self-confidence, self-regulation, and self-awareness
- follow the appropriate stress-management technique
- have good multicultural sensitivity

As per the Education & Training Foundation's expression on the professional standard, the following are the characteristics and behaviors that are to be developed as professional skills:

- Developing expertise and skills to ensure the best outcomes from the learners
- Motivating and inspiring learners to promote achievement and develop their skills to enable their progression
- Planning and delivering effective learning programs for diverse groups or individuals in a safe and inclusive environment to accommodate all types and states of learners
- Promoting the benefits of technology and supporting learners for their better use
- Addressing the mathematics and English needs of learners and doing work creatively to overcome individual barriers to learning
- Enabling learners to share responsibility for their learning and assessment as well as setting goals that stretch and challenges
- Applying appropriate and fair methods of assessment and providing constructive and timely feedback to support progression and productive achievement
- Maintaining and updating teaching and training expertise and vocational skills through collaboration with employers in and out of the campus
- Contributing to organizational development and quality improvement through collaboration with others meaningfully

Teaching as a Profession and Teachers Are Professionals

A qualified teacher is a vital component in the teaching-learning process, irrespective of the modes of instruction. Teaching is a process that facilitates learning and it is the focused application of knowledge, skills, and features designed to offer distinctive services to meet the educational requirements of the individual and society. It is evident that the choice of learning activities whereby the goals of education are realized in the school is the responsibility of the teachers, and hence, teaching claims the development of values among students and guides them for their social relationships where teachers employ practices that develop positive self-concept among the students suitably. In this context, the qualified teacher is the essential element in the delivery of instruction to his students whatever may be the modes of instruction. In this perspective, a teacher should have professional knowledge and skills gained through formal preparation and experiences and teachers provide personal-caring service to their students by diagnosing their needs and by planning, selecting, and using methods and implementing proper evaluation procedures that are designed to promote learning excellently. Therefore, teaching is the process of extracting the expected outcomes among the students, and it includes the following major processes:

- understanding and obeying legal and legislated agendas and policies
- identifying and taking action for students' learning
- offering effective and approachable instruction
- assessing and communicating students' learning
- initiating and maintaining professional relationships among teachers and experts
- engaging himself in innovative professional practices and training

The teaching profession is the queen of all professions, and it is involved with visible, experiencing, and explicating processes. It must give assurance for the holistic development of students through effective teaching strategies concerning content and combinations of students' community. Also, teachers have the right to select the right method, the right approach, and the right mode through which the expected outcome will be possibly drawn without any delay. Hence, to extract excellent output from the students by the teachers, the following are the general and basic skills that are to be imbibed among the teachers without any compromise

- Fluency in the delivery of content
- Mastery in the subject area
- Competitive with co-teachers
- Communicative with other
- Participative in all academic inputs
- Collaborative in producing new output
- Interactive to provide appropriate results
- Technically versatile to meet the changing situation
- Adaptive strategy to satisfy learners' interest and expectations
- Accommodative involvement

- Accountability in all doings
- Accepting the trends in technology and social media
- Learners – continuous learners
- Researcher in his field of study
- Trainee in practicing skills for further action
- Participants to produce good results

Teaching is not a single task to transfer the content from a book to the brain of the students; it is a process to make all-around development among the students.

Techno-Pedagogical Skills among Teachers

Technologies especially information and communication technologies (ICTs) in education have tremendously changed their entire doings and their stakeholders. Of course, technological tools in education have enabled education to an incredible state and status. Integration of different available and applicable technologies in the teaching-learning process made the process more effective and energetic among all levels of learners without any time and distance limitations. Particularly, distance mode instruction or online mode instruction learning could make learning easier and more economical in this present pandemic scenario. In addition to pedagogy, techno-literacy is one of the compulsive competencies among the teaching community not only in this present situation, but it is being practiced and is an expected qualification among the teachers along with their pedagogical efficiency through their qualification degrees for their profession. Hence, it is simply named techno-pedagogy which is not separate or distinct and was used as pedagogy alone before. In the digital era, students are not confined within classroom boundaries; they have a fond interest in learning more learning through other sources or media by using different types of devices or gadgets available to them nowadays. Hence, it is necessary to have techno-based classroom teaching techno-teaching or simply techno-pedagogy as one of the competencies expected to build within teachers or teachers with inbuilt techno-pedagogical skills.

It is a well-known fact that education is considered a systematic process of obtaining skills and knowledge since it is a life-long process through formal or informal exposure, ideas, and experiences. Hence, teaching requires not only the knowledge of content along with knowledge about psychology, philosophy, and sociology of education but also includes technology along with techno-pedagogy with its respective skills to attain its goal without any delay or deviations. Techno-pedagogical skills are nothing but the skills needed to use technology for pedagogical reasons and the competencies to integrate appropriate and effective technologies into teaching. In this digital epoch, techno-skills and digital knowledge are the basic needs of teaching and learning as well. Hence, there is a paradigm shift in the teaching-learning process because of the revolution in ICTs, and hence, teachers are changing their teaching styles, methods, approaches, and instructions to accommodate available effective technologies for their teaching. In this regard, there is no doubt that the use of ICTs in education improves not only the quality of education but also enhances the number of learners and brings out desirable development effective, and productive. The effectiveness of ICT depends not only on competent teachers but also on technologically and technically enriched teachers with sound pedagogy. In this context, teachers at all levels, from pre-primary to university level, need to realize that a range of tools, apps, media, and modes of instruction exists or are to be used for a particular task of content delivery. In simple, in these digital requirements with techno-advanced classrooms, whether face-to-face or online mode, there should be an integration of suitable technologies with effective pedagogy is essential and compulsory to facilitate learning, and this fills the gaps in the educational process not only in the local context and also throughout the world in this current situation, and it should be practiced to have such available e-resources by considering learning difficulties of students in different perspectives to attain our goal in better extent.

Challenges in the Use of Techno-Pedagogical Skills

It is quite natural that the successful integration of technologies and e-resources in pedagogy has amazing benefits and will give a better education for more students with less time and too limitless boundaries, but it also possesses some challenges, and they could be considered and listed below:

- **Lack of training among the teachers and students:** There are no completely trained teachers for the integration of technology in their classroom whether it is a face-to-face or online class. Similarly, students are not trained enough to adapt themselves to avail of such opportunities and resources to a greater extent.
- **Lack of knowledge of information and communication technologies:** Without the appropriate selection of technology and attitude of using such technology for teaching, it is meaningless to have better output. Teachers may not practice techno-pedagogical skills during their course study to the required level and it may reflect in future teaching and they may not cope with the expectations of learners.
- **Poor infrastructures and facilities:** Teachers, who are practicing their skills, are not completely equipped with good infrastructures and facilities to practice and implement their effective professional skills toward their job.
- **No technical support:** Even though the teachers are well-equipped or skilled to have or provide effective techno-pedagogy, they require good technical support to execute their work. The absence of those people may collapse their efforts without any doubt.

- **Lack of institutional support and organizational coordination:** If the new thing is to be integrated, it is required to have full institutional support and co-workers/teachers' coordination. If they fail to take part in the process of initiation of any task, the expected initiation itself will not be possible.

The challenges are not the hurdles to prevent the process of executing techno-pedagogical skills, they could be realized to eradicate them to reach the goal in a better way.

Impact of COVID-19 on Education

The COVID-19 lockdown pandemic situation made everything our doings normal at present. Obliviously, Education and Educational activities have become more and more techno-based and online or remote interaction whatever the nature of work. 'COVID-19 may bring in a paradigm shift in the way professional courses are taught. The adoption of the digital platform to impart teaching will gain significant traction'. A study on 'enrolments in universities that offer both distance and regular modes of education indicates that distance enrolments proportioned about 53% of the total enrolments in FY19. Undergraduate, postgraduate, and certificate courses are in demand for the distance mode by students in remote areas', says a report by *Research Agency India Ratings*. This was done only for higher education, and the status of schools is somewhat different since it is compulsory whether students may or may not go to the schools; they should be alive within that context. In such cases, the impact of COVID-19 made a great change in the work culture of teachers in the schools especially. There is no option among teachers to avail technologies and software for their teaching whether they are paid, half paid, or simply excluded from the routine life of their work. In this pandemic, the levels of teaching may be varied, but the incorporation of technologies by using different software, apps, platforms, etc., are more or less the same by self-instructing and self-learning through different media for their class preparation, presentation, evaluation, and feedback following the instructions being provided by institutions and organization along with guidelines given by the Government. The virtual environment and online classroom may not satisfy the students' expectations and parents' position like emotional considerations, moral development, ethical concentration, dress code, techno-barriers, network problems, and parent's socio-economic status in accessing such classes, but it made its duty to make the learners move on in the educational path to some extent. The government initiatives toward the new education policy and emphasis on research-oriented higher educational institutions/universities are likely to improve the educational standards in India. Also, the government initiatives towards increasing funding decisions for educational institutions through external commercial borrowings and foreign direct investments will positively impact the international positioning of the Indian education sector. Thanks to technology, it made a wonderful change in teachers' lives at all levels, since this situation makes teachers think critically to a greater extent. Of course, Teachers suffered a lot from the pandemic COVID-19, but we as a teacher have never been blocked in all our academic and productive scholastic activities due to this pandemic period. To be frank, comparably teachers are spending their time productively in their daily lives with their students, colleagues, administrators, and office bearers virtually; teachers are preparing their online classes and delivering their content appropriately in different channels, and students are encouraged to engage their classroom activities from their home itself. Here, the important missing dimension is physical touch and emotional consideration. Before this, we people never encouraged our students or simply our children to use mobile phones, laptops, iPods, or any other electronic device for their use, even though they are required for them. Now, the situation changed our mentality and behavior to make use of those gadgets for learning at home. One of the visions of our nation, 'Digital India' has come to its real importance and making use of all the available technology tools for all our routine life in general educational purposes in particular. On the other hand, this lockdown has augmented the adaptation of digital technology; even the person who is not well-equipped in technology moved to avail of remote learning and online practices to a greater extent. This also made teachers' professional as techno-professionals a successful level without any doubt.

Transition of Teaching during COVID-19 Lockdown

This pandemic situation, COVID-19 has made an incredible impact on different perspectives of human lives in general and in education in particular. Due to this unavoidable situation, the world, of higher education has moved to distance and online education by incorporating different strategies, techniques, training among teachers, and purchase of different software with hardware for its effective implementations to a greater level. As per the World Bank (2020), 'Higher education institutions around the world responded by pivoting to distance and online education to reduce or eliminate in-person classes, significantly disrupting the education of what is estimated to be over 220 million post-secondary students or 13% of the global student body. Overnight, there is a wide range of change in education and its dimensions particularly since we are focusing on education in this pandemic situation here. Teachers are being pressured to prepare for online classes with less time without providing appropriate devices for the same since they may not have such technical skills for a larger implementation. When the changes had come in teaching, it was thought that that situation may come to an end within a limited or specified duration, but it continues. There is a different lack – of time to complete the task, resources required to construct the content digitally, technicians' support, techno-skill training and practices among teachers, infrastructures to implement, etc., are to be addressed and rectified for effective implementation of techno-pedagogy for delivery of content with an appropriate methodology for online or distance class teaching. Along with the number of lacking and pressure in the preparation of teachers for online classes, some psychological issues arose among the teachers as well as students and stress among the parents. This pandemic emergency is breeding stress throughout the population and the widespread disease outbreak is associated with unfavorable mental health and adverse psychological issues (Nanjundaswamy et al., 2020; Coronavirus Disease (COVID-19) – World Health Organization, 2020a). Indeed, stress has become a major concern since the COVID-19 outbreak. A study reported that the COVID-19 pandemic has affected the well-being of teachers concerning their profession (Alves et al., 2020). It does not mean that teachers were not stressed before the pandemic, of course, they were/are under pressure and stress in competing the updated content with their competencies with different perspectives. COVID-19 has brought about changes to the education system that impact teachers under the three-gap framework, that is, access, usage, and pedagogical skills. In this situation, it is difficult to reach the students of hard-to-reach and they are really economically weaker sections and are far away from the educational boundaries and are interested in interacting with the educational environment that is affected by this pandemic. It is accepted that all the teachers were not full-fledged techno-literates and technically sound enough to handle such a situation to compensate for the classes online by utilizing all available e-resources and apps to a greater degree.

It is a well-known truth that there is a digital gap that exists at learners' level and teachers' level not only in our country, India, but also across the globe, and this gap range varies from region to region depending upon the infrastructures, the socio-economic backgrounds of stakeholders and many other local factors – culture, customs, etc., As the spread of coronavirus disruptive impact on all lives in the globe, the mode of shopping, entertainment, work-culture, medical consultations, academic events, etc., have moved to online mode. From primary education to higher education, online-based inputs are being provided and practiced to compensate for such a natural educational climate as an alternate way to connect with the learners to contact with educational activities without making them deviate from their actual learning. Hence, different stakeholders – government and private organizations have been trying their best to assist each other by revamp-up their existing online platforms, web applications, modules, content, apps, media, open educational resources, etc., and providing appropriate training to the teachers to use those apps and platforms for their teaching by the level of their learners. Moreover, efforts are being made both by the government and non-government organizations to support the educational institutions where they are required to make a smooth evolution from face-to-face to the virtual world to challenge the situation and to channel them productively. In this context, up-gradation in their teaching skills especially techno-skills in this scenario, and motivating the teaching community in organizing counseling sessions are to be done by the administrators and organizations for its effective execution of strategies in implementing the required classroom environment in this scenario prolifically.

Values Development among the Students

It is a well-known fact that developing moral values among students makes them a good personality since moral values are the building blocks of everyone's personality. Providing education not only makes the students intellectually vibrant but also should give them space to develop good values along with subjects and through the subjects. Values help them to have a good attitude, meaningful beliefs, and productive ideas to develop their personality by understanding the people around them. Teachers' role is not only confining the completion of syllabi, the conduct of examinations, evaluating their assignments, giving feedback to make them get more score further, and also there should be meaningful teaching, that is, teaching is not only focusing on the content prescribed for their level and also it should have values while making them understand the subject, framing the questions by considering all the components of development, assessing the absolute all-round expectations, providing meaningful criticism with space for correction and improvisation further. It is proved that students are very good observers and imitators; hence, teachers are to act as effective role models which could be attained only through meaningful professionalism. Knowing the facts and understanding the context is not only important but also there should be a productive integration of all available technologies effectively inappropriate with the content of the subject and level of the students without deviating from the ethical limit and concentrating the development of the value among them simultaneously.

Need of Student Empowerment Today

Student empowerment is defined as where the students gain the authority and agency to make decisions on their own and implement their changes in the learning context and it includes learning, teaching, and leadership in the classroom. The concept of schooling or teaching is not meant that the process of transferring content from book to brain, then to paper and noted as a number in their rank-cards. It is the place to give space for developing their self-confidence, sense of self-efficacy, ability to solve problems, analyze the given/received information, communicate the people through an effective medium, make use of resources meaningfully, adapt all aversive situations and make them able to face the real-life situation. School is the miniature of society and the teacher is the role model to make them realize the value of his/her role in and out of school effectively. In this, it is also important to consider all the aspects like psychological, emotional, intellectual, economic, and social components. In this, social motivation focuses on the intra-personal and interpersonal needs of students within the classroom and how they translate these needs into social goals. Establishing an empowering classroom environment may facilitate students to fulfill their need for power and, indirectly, belonging, particularly because a need for belonging is linked with a need for power (Baumeister & Leary, 1995). Empowerment is about students satisfying their need for power so that they gain a sense of power with peers or the teacher rather than gaining power over them. Students who are involved in power-with relations are more likely to feel a sense of belonging than if they are involved in power-over relations. In this current situation, students are competent in operating all types of devices for their teaching-learning process, but there is a question 'What about their handling part?' Empowerment is not confined to the concept above mentioned, it also involves the competency of handling and facing not only the persons around them but also everything in their educational process. It is also in the hands of teachers with effective guiding approaches in such aspects to help them reach the real goal of their education as well as their lives.

Conclusion

Nelson Mandela rightly said that 'Education is the most powerful weapon you can use to change the world'. In this digital interactive scenario, education has proven its place and mode through the trends of the day. In the present situation, a teacher is one of the sources of knowledge since students have practiced accessing online materials through online sources, especially after this pandemic. Learning content-wise students were very much improvised but they missed a few important requirements – self-discipline, social interaction, contacting teachers, lack in time management, and p handling of technologies unwontedly. Simultaneously, teachers are to maintain the following to meet their learners productively – technologically sound, updating the trends in techno-teaching devices/platforms/apps, meaningful evaluation, effective feedback, collaboration with other disciplines, and making them involved in techno-curricular activities. It is realized that education without effective teaching by incorporating current trends is meaningless and it is the major part of the process of teaching-learning by considering all the components of this digital world and its expectations. Hence, the educational system around the world requires teachers with a high level of quality and competencies in general e-competencies in particular, and the training of teachers with these qualities who are to be equipped more with professional skills to improve the quality of education. In this new-normal situation given by the COVID-19 pandemic, it is required to be aware of tools, apps, media, and modes for teaching, attitude to accept those available resources, utilize appropriate e-resources for effective teaching and adapt to this digital environment by avoiding the reasons not to be used those things for their teaching because of their age, locality, convenient, environment, exposure, so on. To enhance and enrich the professional skills of prospective teachers whatever may be their level of teaching, it is required to take into account the changing roles and professional skills needs of teachers in the present digital scenario. The professional skills for this situation are not only equipped in different perspectives for existing teachers but also for the practicing teachers during the study in a deep manner by realizing the situations and expectations of upcoming learners to compete with the world-class teaching and to meet the challenge of the digital world effectively.

References

Alves, R.; Lopes, T.; Precioso, J. (2020). Teachers' Well-Being in Times of Covid-19 Pandemic: Factors that Explain Professional Well-Being. *International Journal of Educational Research and Innovation*, v15, p203–217. https://doi.org/10.46661/ijeri.5120

Baumeister, R. F.; Leary, M. R. (1995). The Need to Belong: Desire for Interpersonal Attachments as a Fundamental Human Motivation. *Psychological Bulletin*, v117, n3, p497–529. https://doi.org/10.1037/0033-2909.117.3.497

Doyle, A. (2024, May 10). *Top Soft Skills Employers Value with Examples*. The Balance. https://www.thebalancemoney.com/list-of-soft-skills-2063770

Nanjundaswamy, M.H.; Shiva, L.; Desai, G.; Ganjekar, S.; Kishore, T.; Ram, U.; Satyanarayana V., et al. (2020). COVID-19-Related Anxiety and Concerns Expressed by Pregnant and Postpartum Women-A Survey among Obstetricians. *Arch Womens Ment Health*, v23, p787–90.

World Bank (2020). *The COVID-19 Crisis Response: Supporting Tertiary Education for Continuity, Adaptation, and Innovation (English)*, pp. 1–9. Washington, DC: World Bank, 2020 Group. http://documents.worldbank.org/curated/en/621991586463915490/The-COVID-19-Crisis-Response-Supporting-Tertiary-Education-for-Continuity-Adaptation-and-Innovation

World Health Organization (2020a). *Coronavirus Disease (COVID-19) Pandemic 2019*. Accessed Jan 30, 2020. [Online]. Available: https://www.who.int/emergencies/diseases/novel-coronavirus-2019

Chapter 2

Cognitive Learning

Hima Bindu Maringanti and Meenakhi Sahu

Learning Outcomes

- This chapter would help teachers evaluate students based on their learning style and map them onto one of the known theories.
- Upon understanding the student type, they may be clustered into groups and customized teaching methodology could be adopted by the teacher.
- Students, who read this chapter, also would be able to self-evaluate and identify themselves to fit into one or more of the learner types.
- Students could also try to adapt themselves to a different learner type if they find it interesting or easier to learn.
- CLA could be explored and learning skill enhancement can happen.

Prologue

The objective of this chapter on Cognitive Learning is to elucidate the definition and its genesis; why and how this concept started and what are the implications. Cognitive Learning is the ability of the brain to enable learning by attention – sensor-level capture of stimuli from the environment, selection-focusing (through sensor-level filtration process), chunking into related concepts, thinking about own thoughts and question why and how (termed as cognitive understanding), understanding the underlying concepts and principles, solving problems with the understanding so achieved and responding to the stimuli (output generation to the environment). Not all students have the same IQ (intelligence quotient) and hence their level of understanding of things to learn also varies. Each student has his or her way of learning and also solving problems, including speed and efficiency.

DOI: 10.1201/9781003504894-2

Later on comes the issue of memory storage and recall at the appropriate time and in the expected duration. Many factors influence this functionality; some are known to be cultivatable, while others are intrinsic and very rarely cultivated.

Introduction

Cognitive Learning (CL) is the involvement of the brain in the process of learning. Out of the nine cognitive skills a human has, six of the skills, namely, attend, listen, think, understand, solve, and respond, are included in CL. The concept of CL applies both to the teachers and students in the Teaching-Learning process, which is an important parameter even in NAAC's A&A process. Not all students have the same IQ (intelligence quotient) and hence their level of understanding of things to be learned, also varies. Each student has his or her way of learning and also solving problems, including the speed at which they learn/solve and efficiency (level of understanding – partial or complete, even quite different interpretation of the subject matter).

The external stimuli could be verbal or non-verbal and visual or auditory. Accordingly, the information processing and their internal representation also would be different. Suppose it is a verbal stimulus(say tree). In that case, it may elicit a specific area of memory containing the visual image of a tree, green in color, big or small in size, flowering or fruit-filled, etc., or may elicit an emotion if the stimulus is a reinforcer, positive or negative (an appreciation or a scolding).

These are termed representational associations, verbal or non-verbal of the stimuli with the cognitive structures of the mind [1]. Two types of memory are involved here, the working memory (Short-Term Memory), where processing happens, and the permanent memory (Long-Term Memory), where it is retained for future recall and use. The retention in STM is very short, of the order of 20–30 seconds, unless refreshed within that duration, while the LTM stores permanently, ranging for months, years, or decades, in the form of chunks of memory consisting of related items in the form of an associated concept map or brain map. The crux of this memory is the recall efficacy, which is largely dependent upon the refresh rate, that is, how often it is remembered. But sleep deprivation would affect the recall efficiency and trauma – physical or mental would erase parts (usually past events closer to the time of occurrence of the trauma) of the LTM and keep the very old information events intact. Cognitive psychology is the study of one's internal processes and is closely related to CL.

Definition

CL is the effective involvement of the mind in the learning process, using six of the nine cognitive skills in the process. This learning merges two processes: cognition and learning.

Cognition is the mental process of gaining knowledge through the understanding of external stimuli, thinking, and experiences over time. According to CL theory, certain thinking processes can lead to more knowledge retention.

Any person adopting CL could thus become successful in his/her domain, student, teacher, or any other professional. The prime elements of CL are memory, comprehension, and application, and they largely focus on learning from the past, applying the experience to the present problem, and retaining the solution process in the memory so that the retention is longer. This is termed *Classical Conditioning* [2], by which the mind remembers the *smell of a rose* or the *ringing of the bell* as associations between concepts; the sweet smell is associated with that of a rose, and the ringing of the bell is associated with meal time. Addictions are affected by classical conditioning, as all events related to a specific 'smell' or 'ringing bell' are remembered and make one crave for something that was quit. *Operant Conditioning* [3] is the other form of associative learning, wherein one's behavior (learning) is dependent upon reinforcement or punishment and causes repetition in the future. If one is applauded for a specific achievement, or a problem solution, for example, the student tends to repeat (exhibit) the same or similar behavior (learning) in the future.

Genesis

Jean Piaget's theory of Cognitive Development [4] in the early 1950s and Jerome Bruner's [5] work focus on how mental processes are linked to teaching, followed by Vygotsky [6] are some of the significant contributions in the direction of CL.

The question of whether the learning abilities of children are inborn or could be cultivated and enhanced with certain training was answered by these works. Social cognitive theory (SCT) and Cognitive behavioral theory (CBT) [8] are the two categories of CL theory in the psychology domain.

While the SCT is based upon observation of others and adapting self accordingly, the latter is based upon mental processes such as thoughts and interpretation of life's events. For the benefit of the individuals and the competitive world, at large, CL theory was discovered resulting in long-term benefits like learning enhancement, boosted confidence, increased comprehension, improved problem-solving skills, learning new things faster, and abstract thinking/concept formation. Various CL strategies like Learner-centered, Meaningful experiences, Learning through Discovery, and Personalized Learning have been shaped by various psychologists including Piaget for implementation in the corporate world. However, these strategies are very well suited to the education field.

Types

Not all students have the same IQ (intelligence quotient) and hence their level of understanding of things to be learned also varies. Each student has his or her way of learning including speed and efficiency.

The various learning types are Implicit learning, Explicit learning, Meaningful learning, Cooperative and collaborative learning, Discovery learning, Non-associative learning (habituation and sensitization), Emotional learning, Experiential learning, Receptive learning, and Observation learning [9].

Implicit learning does not involve an active intention to learn, but the skill is inherently acquired without knowledge of self, like walking, talking, applying breaks on a well-known road, etc.

Explicit learning is consciously intending to acquire new knowledge with continuity in action and sustained attention, like attending a workshop on Communicative skills to enhance your skills.

Meaningful learning is a means of improving your academic/managerial skills for better performance, like attending a week-long hands-on workshop on Machine Learning.

Cooperative and collaborative learning includes group/team learning as the sum of individual inputs is lesser than that of the group as a whole. Group interaction, reinforcing interdependence amongst members, individual responsibility, and equal participation are its elements.

Discovery learning involves deep thinking and research in a domain that is not the primary field of research, involving adapting to the new technology.

Non-associative learning includes habituation – learning by habit due to prolonged exposure to stimuli with reduced reaction and sensitization – sensitivity to stimuli, specific to your focus and interest, like setting different caller tunes for important phone numbers to increase sensitivity.

Emotional learning is understanding the emotions of self and others and moderating them, enabling balanced interpersonal relationships with different people.

Experiential learning is the way of learning through experiencing situations. However, the same experience by two different people may have different results, depending upon their perception and amount of introspection and self-control.

Receptive learning is the one-way general classroom learning method, where one just listens and accepts whatever is said, without any interaction.

Observation learning is the technique of learning by observing others, whom you feel are good masters of their field, either academics or administration, and aspire to become like the role model or sometimes excel and surpass the model.

All the above types of learning may either be the addition of new knowledge to the existing, knowledge corroborating (in harmony/agreement) the existing knowledge, or contradicting the existing knowledge. These are termed *accretion* (adding a new fact to an existing schema), *tuning* (changing the existing schema so that it becomes

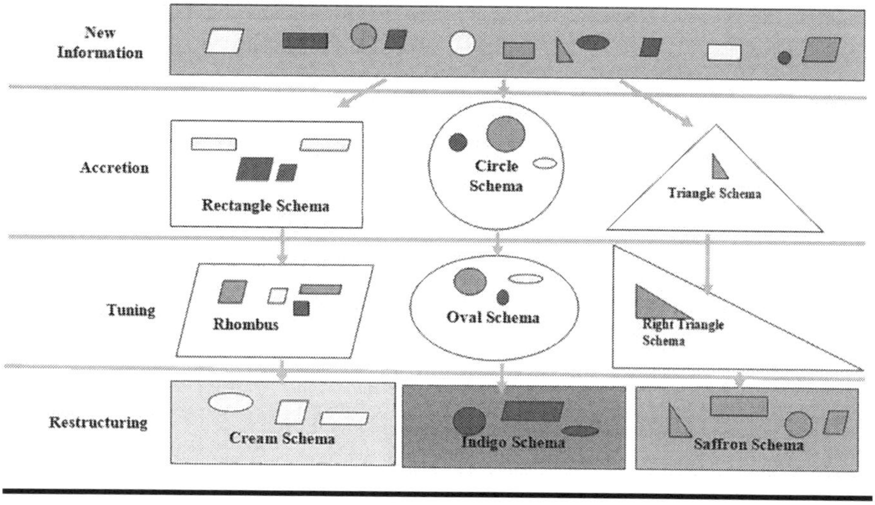

Figure 2.1 Schema theory. Credit: Obsidian Learning [10].

more consistent with experience), and *restructuring* (creating an entirely new schema that replaces or incorporates old ones) [10], as represented in Figure 2.1.

Classical Theories of Learning

Learning is the process of performing the symbolic representations to memory where they may be processed. The study of learning is done through the study of memory. There are three important cognitive theories for learning. Those are Piaget's developmental theory, Lev Vygotsky's socio-cultural cognitive theory, and George Miller's information process theory.

Jean Piaget [4] concerns the learning ability of children. According to him, a child can comprehend the outer world through a sequence of four discrete stages. The cognitive development is a sequence of structural transformation. The stages are sensorimotor intelligence (birth to 2 years), preoperational thinking (2–7 years), concrete operational thinking (7–11 years), and formal operational thinking (11 to adults). Each stage involves a different type of intelligence.

The child's thinking at each stage of development is different from the other stages. The learning of the child from birth to two years is known as the sensorimotor stage. The infants understand the world through senses and movements. The objects that are in front of the infants are recognized. Sensorimotor development happens in this stage. Without knowing any language the child understands the existence of objects and expresses it with motor actions. Next is the preoperational stage, the age from two to seven years.

At this stage, children begin to think symbolically and learn to use words and pictures to represent objects. When the child begins to use symbols, the cognitive structure develops. This structure will be used in the future and also may help with the construction of a new or developed structure. While playing, children deploy the objects in various new activities. They do not think rationally. Meta-cognition develops in this stage. For that reason, play school teachers express themselves through activities. At this stage, children acquire conversation skills. The concrete operational stage is the age from 7 to 11 years. During this time, children redesign their thoughts more smoothly and logically. Thus, children of this age are taught arithmetic tasks. The child becomes aware of her environment and becomes less self-centered. Moreover, they have enhanced reversible ability in the preoperational stage. Piaget stated that the three basic reasoning skills acquired during this stage were identity, compensation, and reversibility.

The last state is a formal operational stage, the age from 11 onwards. Children of this age can think abstractly and achieve skills such as inductive and deductive reasoning abilities. The child becomes able to reason not only about tangible objects and events but also about hypothetical or abstract ones. According to Piaget, each child goes through the stages in the same order. Piaget claimed that knowledge cannot simply emerge from sensory experience; some initial structure is necessary to make sense of the world.

Lev Vygotsky's [6] stated social cultural cognitive theory. The theory not only focuses on how adults and peers influence individual learning but also on how cultural beliefs and attitudes affect the learning process. The connection between language and thinking goes through several transformations in the course of cognitive development. The theory is comprised of concepts such as culture-specific tools, private speech, and the Zone of Proximal Development.

Socio-cultural theory is a theory of the development of higher mental practices. The theory suggests that human learning is largely a social process. The mind of infants is composed of biological constraints. And the change of structure happens with the learning. Each culture provides tools for intellectual adaptation. These allow children's brains to become adaptive to the culture in which they live. Social interaction is the core of the communication and learning process of a child.

According to this theory, the potential for cognitive development is limited to a zone of proximal development. Children should be taught considering this, as the grasping and understanding capability of individuals differs with zone. Scaffolding is when a person can almost perform a task but needs support from peers, adults, teachers, and other mentors; learning occurs through interactions. The psychological structures do not exist in the individual's mind rather they are formed as a result of interaction with the social context. Humans learn from their communities. Scaffolding competence assistance is usually provided through mediation of the environment by a parent or teacher, by which cognitive, socio-emotional, and behavioral forms of development are facilitated. The concepts of socio-cultural theory are the culture is imperative in learning, language is the core of culture and the persons learn and develop from their activity in the community. Collaborative learning, discourse, modeling, and scaffolding are strategies for supporting the intellectual knowledge and skills of learners and facilitating intentional learning.

Socio-cultural cognitive theory emphasizes how social factors influence development. Vygotsky emphasized that cognitive development can differ between different cultures. Whereas Piaget's theory focuses on how a child's interactions and explorations influence development, his theory suggests that development is mainly universal, whereas, as per socio-cultural theory the development happens locally.

He considered that brain structural change is necessary for learning. Bruner was concerned with how knowledge is represented and organized through different modes of thinking. Bruner's research of intellectual development proposed three modes of representation. However, the modes are not age-dependent as in Piaget's stages of developmental theory. According to Bruner, the modes are not strictly age-dependent. The three modes of Bruner's theory are – enactive, iconic, and symbolic.

Enactive Representation (Action-Based)

The enactive stage appears first and happens in the very young generally from childbirth to age one. This stage involves the encoding of information and storing it. Infants learn by action, rather than by thinking. For the child, the thinking is based entirely on physical actions. The action-based information is encoded into knowledge representation. Then, the internal representation is stored in the memory. There is a direct approach of objects; without any pre-representation of that. In this stage, the knowledge is represented mainly in the form of motor responses.

Iconic Representation (Image-Based)

This type of representation happens in children between one and six years old. Visual information is stored in the form of images. In this stage, external objects are internally represented in the form of mental images. The knowledge is the visual summarization of images.

Symbolic Representation (Language-Based)

This type of representation happens in children above seven years. In this stage, knowledge is primarily stored in the form of words or symbols, as in any language scheme. Each symbol has a predefined relation to something it represents. The words and symbols are combined to describe knowledge. Iconic representation develops an ability to think in abstract terms. This mode develops at last.

Both Bruner and Vygotsky emphasized a child's environment, especially the social environment, more than Piaget did. Both agree that adults should play an active role in assisting the child's learning. In addition to the above three classical theories, George Miller's Information Processing Theory [7] also found a place in literature. Information processing is a cognitive theory that focuses on how information is encoded into human memory. The processing of the human brain is compared to processing through the computer.

The Information Processing model explains the process of information perceived through the sensors, encoding, storing, retrieving, and making use of the human brain. It is hypothesized that processing involves three stages: encoding (collecting and representing information), storage (holding information), and retrieval (obtaining the information when needed); a control process that determines how and when information will flow through the system. The processing is analogous to a computer model. The computer reads data through input devices, decodes and processes the data, the processed results are stored in secondary memory for future use otherwise discarded retrieves the stored information for use as and when required. Similarly, the mind takes information, performs operations on it, changes its form, stores processed information, retrieves it when needed, and generates responses to it.

Sensory memory is the first stage of Information Processing Theory. The brain perceives the environment through the sensory organs, viz., eye, ear, skin, mouth, and nose. The input stimuli are visual, hearing, touch, taste, and smell. Humans selectively attend to one activity of his/her interest out of the available and tend to ignore other stimulations. The attention can be distracted by something else too. Information is filtered from our sensory memory into our short-term or working memory.

Some of the information we hold in our short-term memory is discarded or filtered away once again. Short-term memory remembers things for very short periods. And the size is also very small. The processing of short-term memory is affected by factors such as individual cognitive abilities, storage capacity, and interest in self. Then selective processing is used to bring our attention to those details so that we remember them for the future.

Repetition is used to transfer vital and fascinating information from short-term memory into long-term memory. A portion of the information is stored in long-term memory. Long-term memory is stored permanently in human brains. There are three categories of long-term memory: Semantic Memory used for verbal information or meaning. These memories are stored as propositions, images, and schemata. Next is Episodic Memory, which stores events and information related to a particular place and time. And third type is Procedural Memory which stores the intelligent actions and the procedures for 'how to act'.

More time is required to develop procedural knowledge, but once learned, the knowledge can be remembered for a long time in the long-term memory.

Plato and Descartes [11] were the first philosophers who focused on cognition and tried to study about human thinking process. Later on, many others delved deeper into thinking. Jean Piaget is the father of cognitive psychology; his work focuses on the external environment and internal brain structure, to study their impact on learning.

Cognitive theory has developed over time but with sub-divisions of learning and understanding. At a rudimentary level, cognitive theory has two important contributory constituents, viz., internal thoughts and external forces. To understand this and to find out means of analyzing the impact of their thinking on learning and behavior becomes the utmost responsibility of a student; who then tries to master control of his/her thinking.

Now it is the turn of the teachers who should give opportunities to students to ask questions, think out loud, and maybe fail at times. These coping strategies can help their thought process to utilize this knowledge to fabricate optimum learning opportunities.

Contributory Elucidation

Objective and Concept elucidation

The objective of this chapter on CL is to elucidate the definition and its genesis, why and how this concept started and what are the implications.

CL is the ability of the brain to enable learning by attention – sensor-level capture of stimuli from the environment, selection-focusing (through sensor-level filtration process), chunking into related concepts, thinking about your thoughts (termed as cognitive understanding), understanding the underlying concepts and principles, solving problems with the understanding so achieved and responding to the stimuli (output generation to the environment), as shown in Figure 2.2.

The limitation of the learner is due to the capacity of the STM and its retention period; thereby he/she uses various strategies to select and store relevant information. So, it is now in the hands of the teacher, the *Instructional Designer* [1], to use different techniques to enhance learning by simplifying the learner's assimilation of information into their schemas [10]. Relating the existing knowledge to the present by way of *metaphors* or *problem-reduction* strategy in AI also is a strategy used

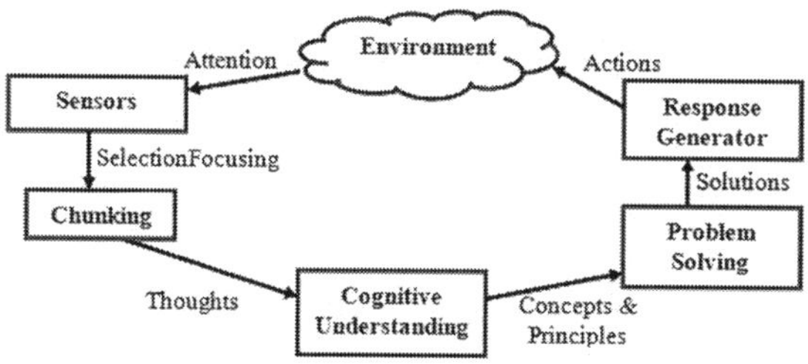

Figure 2.2 Functional Block Diagram of Cognitive Learning.

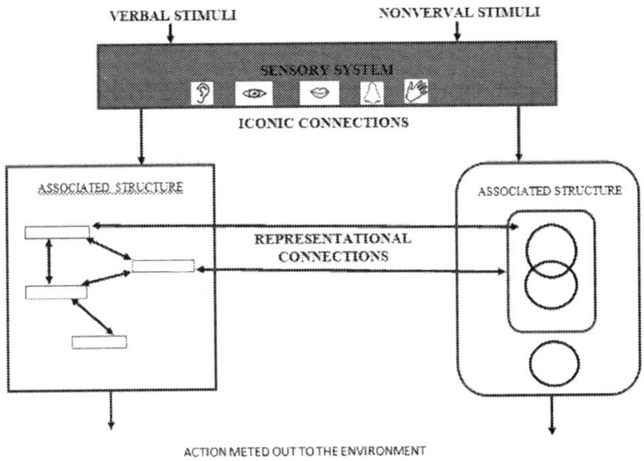

Figure 2.3 Dual Coding Theory [13].

to enhance retention. The Dual Coding theory (DCT) [12] suggests verbal and non-verbal stimuli and their associations as mechanisms of information representation, as shown in Figure 2.3. He claims that applying DCT to multimedia instruction for learning has been justified as the capability to remember pictures better than concrete words is estimated at a ratio of 2:1; hence it is by engaging both audio and visual channels (as video does) that learning is most effective.

In addition to DCT, another theory called as the cognitive load theory (CLT) has been proposed, which suggests ways in which instruction can be designed to maximize the efficacy of STM by minimizing extraneous or distracting information. Learners are thus enabled to process and store relevant knowledge in LTM.

There are three types of cognitive load [13,14], namely, *Intrinsic load*-difficulty that is inherent in the content itself; *Extraneous load* – information or activity that is not needed for learning and *Germane load* – cognitive processing that assists in the formation of learner schemas. Figure 2.4 below shows the mechanism of Obsidian Learning [15] and CL distribution and the effects of instruction design on each of these three loads.

The third theory is the Cognitive theory of multimedia learning [16], similar to the three categories of CL, but called Extraneous, Essential, and Generative processing, respectively. In addition, his theory also includes 12 principles of multimedia learning to facilitate the above three types of processing, viz., coherence, signaling, redundancy, spatial ambiguity, and temporal contiguity, as listed and described in Table 2.1.

Also, student user profiling can be done to capture their web behavior and the e-content customized to suit their requirement using AI techniques. Learning through fun and games, problem-solving, fieldwork, collaborative teamwork projects, etc., would enhance the learning skills of students.

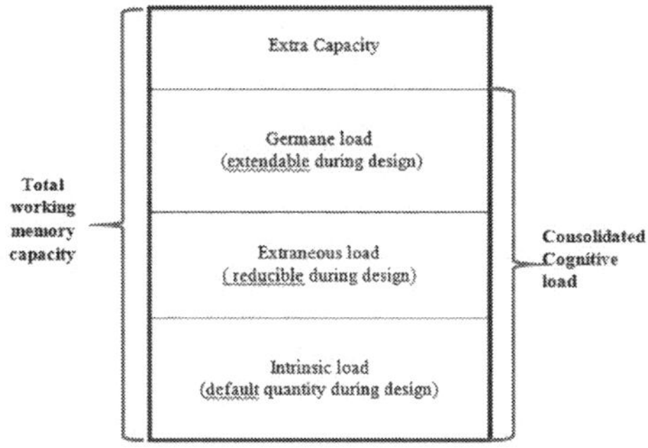

Figure 2.4 Cognitive load theory model. Adapted from Moreno & Park, 2010. Credit: Obsidian Learning, [13, 15].

Table 2.1 Principles for Managing Extraneous Processing [16]

Principle	*Description*
Coherence	Learning is improved when extraneous material is removed. Extraneous material includes interesting but irrelevant words, symbols, pictures, sounds, and music.
Signaling	We learn better when visual or textual cues that highlight content organization are included. These cues assist the learner in schema development.
Redundancy	We learn better from graphics and narration than from graphics, narration, and printed text. The visual channel can be overloaded w the learner must process both text and graphics.
Spatial contiguity	We learn better when related words and pictures appear near each other because the learner does not have to use cognitive resources to put related elements together.
Temporal contiguity	We learn better when related words and pictures are presented simultaneously rather than successively. Presenting related items simultaneously allows the learner to hold mental representations of both in WM together.

The instructor needs to choose/develop his/her strategy to teach a particular subject and implement it in the classroom so that most of the students benefit from this. He/she also needs to be aware of the various learning theories existing, including the techniques for teaching and classroom management, and then apply them in the classroom. Understanding the learning theories will also help teachers connect to all different kinds of students and hence cater to their varying learning needs. Teachers can focus on different learning styles to reach students, developing teaching aids that focus directly on student needs and aptitudes.

Alternative techniques of enhancing cognitive abilities including Learning comprise conventional techniques of physical and mental well-being. These include Yoga and Meditation (Y&M), Sudarshan Kriya Yoga (SKY), Chakra/Mudra Pranayama, Sahaj Samadhi meditation, Sudarshan Chakra Kriya, Mindfulness meditation which are scientifically proven to enhance concentration, comprehension, efficiency of performing tasks and reducing anxiety and stress.

The impact of these techniques on the psyche of participants and the learning enhancement was evident by looking at the higher grades secured by students as well as teachers; even addictions were corrected. The students learn to take responsibility and become mindful of their goals and means of achieving them with perseverance, integrity, and confidence.

Impact Evaluation

The different steps to be adopted for evaluation of CL are the following:

- Focusing on the educational domain, **pre-testing** of students' learning abilities and learner types needs to be the primary step to be carried out in all classes by all teachers in all subjects. The interest in learning the various subjects is certainly going to be different and the teaching methodology adopted by the teacher also is an important parameter that impacts the students' learning interest. *Formative assessment scales* could be used to assess them for learning.
- **Segregating** the students into clusters based upon the assessment carried out in step 1, based upon learning motivation, learner type, and skill requirement.
- Identifying and **tagging** the clusters with Learning Type.
- Choose a **methodology** of teaching for each cluster and continue imparting teaching for a chosen number of hours/days/months/years.
- Learners' learning achievement could be **tested** using regular (continuous) tests requiring problem-solving skills with incrementally increasing difficulty and based on the application of theoretical concepts in problem domains using *Summative assessment scales* for assessment of learning.
- Provide **feedback** to students about their performance in the step above and also give insights into the future course of action for their learning and take responsibility.
- Design a Questionnaire to take the feedback or **self-assessment** from students about their opinion of achieving learning targets and use these for validation purposes.
- Any large errors or **discrepancies** may be considered for revision and re-incorporated into this **cyclic** process of evaluation.

Limitations

The context and situation of students like their family background, their living conditions, exposure to digital media, approachability to educational institutions, health condition – both physical and mental, food and nutrition, social status, medium of instruction and hence comprehension, are some of the other parameters, all of which cannot be changed for good. Anyway, these parameters have not been included for discussion here. Those characteristics that are changeable have only been addressed.

Applications

Some of the application domains of CL are described below, but are not limited to these.

Education

The teaching–learning process is the heart of education in addition to progression to higher education, placement, and excellence in career and life. In the formative years of a child's education, she is tested for the level of intelligence using many scales like Wechsler's, Binet's, and Terman Merril's, the major constituents of which are verbal comprehension, numeric, spatial, and logical reasoning, and pattern matching. Physical classroom teaching has its advantages and disadvantages; the teacher and the students are in proximity and body language and eye-reading help the teacher gauge her students' understanding level.

So, the psychological satisfaction between the teacher and students is an important factor that influences learning. Digital learning enables anytime, anywhere learning, enabling students to learn according to their choice from the comfort of their home/bed/living room, provided they are digitally connected.

The learning management system (LMS) is an option for the students and teachers of modern times, especially during the pandemic. A full-fledged LMS, populated adequately with study material embellished with multimedia content, is a boon to students with different cognitive abilities and learner types. The modern student learns to adapt herself to the demanding current situation and the dynamically changing technologies, to keep himself/herself abreast in the job field, well equipped with enhanced skills, may it be programming languages, communication, and analysis tools.

There exist five educational learning theories, each with an insight for teachers/students, to adopt different teaching techniques, catering to the needs of the differing learning needs of his/her students. They are mentioned below:

Cognitive Learning Theory

The CLT enhances student learning through a process of self-analysis of their thought process and hence moderating it to achieve optimum results. Teachers should give opportunities for to students ask questions, and think out loud, and may fail at times. These coping strategies can help their thought process to utilize this knowledge to fabricate optimum learning opportunities.

Behaviorist Learning Theory

Reinforcement is a method of teaching pets to be disciplined; also be used for training students; akin to the age-old 'carrot and stick' strategy. 'Good', 'well done', 'you can do better', etc. are some of the positive reinforcers that are or can be used by teachers to encourage them to learn better and also retain longer; a direct outcome of behaviorist theory.

Constructivist Learning Theory

Constructivist Theory is based upon the concept of learning through experience earned in the past. Such students just append the current knowledge to the previously experienced knowledge and thus create a new model of learning, unique to them. Hence this type of learning is categorized as an active learning process, moderated to suit each learner. In such situations, teachers act as mere guides observing and evaluating the learners to create their understanding of things and concepts

Humanism Learning Theory

This theory relies on creating an environment for students, where they realize that all their basic needs and necessities are satisfied and so, feel contented, but for a short while. To create such memorable moments, teachers and educators have to take the responsibility of providing the basic facilities for a problem-free learning environment; for example comfortable seating, illumination, library, laboratories, canteen, sports, washrooms, hostel, etc.

Connectivism Learning Theory

As the name implies, this theory is based on a collaborative and group-based study, projects, assignments, presentations, etc., so that students learn to cooperate, rather than compete, and also make the work enjoyable. Clubs, Research groups, Project groups, SIGs [Special Interest Groups]. So, the responsibility of the teachers here is to enable grouping and assign group assignments according to their interest area, to make the experience pleasurable and not a burden.

In addition to the above five theories, three other theories for the benefit of students are mentioned below, to be adopted by the teachers. They are

Transformative Learning Theory

The founder of this theory is Jack Mezirow, who believed that self-analysis and introspection help adult students to change themselves for the better and hence their understanding capability. This strategy helps students adopt rational thinking, rather than believe and follow superstitions. Also, the students will be able to make out the difference between good and bad; correct and incorrect; ethical and unethical, etc. Teachers could leave open a for seeking opinions and group discussions to gauge their creativity and perception of things.

Social Learning Theory

Albert Bandura was the founder of this learning theory, in which he conducted the Bobo doll experiment, to teach children, social behavior and exhibit politeness or aggression in a given situation; in the process learn by observation and assessment, of what is good and/or bad. Attention, retention, reproduction, and motivation are the important constituents of this theory and the teacher guides the students accordingly.

Experiential Learning Theory

David Kolb first identified this theory of 'learning by doing' or experiential learning. Research shows that anything that is experienced is remembered longer and also retrieved later effectively; rather than just reading. Questions and tests involving the application of theoretical concepts find a primary place here for effective learning, retention, and recall. Group learning enabled through connectivism also encourages this type of learning, added with motivation.

User Interface Design

A user interface is evaluated for its ease of use and flexibility; in addition to minimal surprise while interacting with the system through it. Commonly used, but well-designed interfaces include the ATM, mobile phone interface, and the Amazon portal. No one needs to teach us or train us on using these; the reason being they are self-explanatory and intuitive, with expected responses and the flexibility of undoing actions taken by mistake. CL is categorized into five levels, viz., remembering knowledge, understanding, application, analysis, evaluation, and creation [17]. While knowledge leads to understanding, understanding leads to application. An effective user interface impacts a change in the behavior of the user through rewards (positive reinforcement) or punishments (negative reinforcement); for example, a good move or action is appreciated by a reward that may be verbal, visual, or audio with clapping sound or a reduction in points

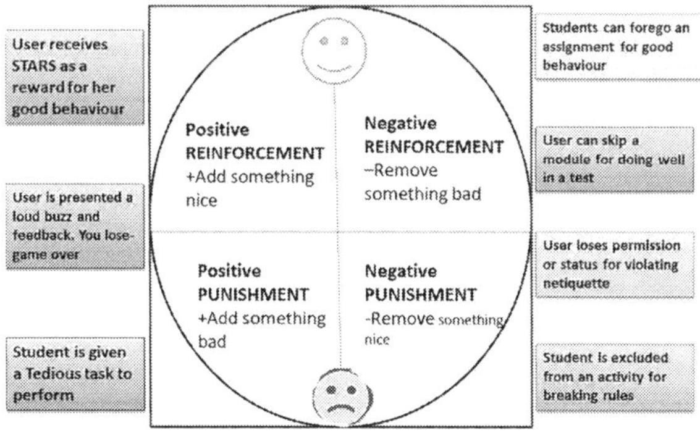

Figure 2.5 Behaviorist Learning Strategies with Reinforcement [3].

to be scored in a game because of the delay or a wrong move with a red colored remark.

There are two types of users: one who does not show behavior on receiving a punishment, but shows behavior only when appreciated; the other who is motivated both by positive and negative reinforcement (Figure 2.5), to perform better, so changes behavior accordingly, that is, learns from mistakes.

As per *Thorndike's Law of Effect*, behaviors are associated with pleasure and comfort and are very likely to be repeated and those behaviors associated with pain and displeasure are not. Games and gamification are highly dependent upon rewards and children who play online games grow up to be confident individuals and learn to take challenges, to get these rewards again and again.

Interface design in learning environments involves instructional design as envisaged by Benjamin Bloom in 1950 involving three taxonomies, viz., cognitive, affective, and motor-sensory. Out of these, the cognitive taxonomy is mentioned here, which is classified into five categories: knowledge, comprehension, application, analysis, synthesis, and evaluation, arranged in order of lower to higher cognitive complexity, revised later on in 2001 shown in Figure 2.6.

Other applications include the **Corporate world**: following majorly the Skinner model of reward and punishment with incentives and perks on performance and stopping of increments, cancel leave, etc. as punishments. According to performance and skills, employees are assigned or placed in appropriate levels according to Bloom's taxonomy.

Nursing is the other domain where CL could be applied directly, based on skill training and experience the nurses are placed at appropriate levels and departments or specializations, like cardio, physio, gynaec, urology, radiology, and ICU.

Experiential learning is an asset for them as they tend to learn and acquire new practical knowledge while attending to patients. Markets, either physical or electronic are based on customer/**consumer behavior** and demand. Consumers tend to usually go by rating mechanisms or recommendations of friends and other family

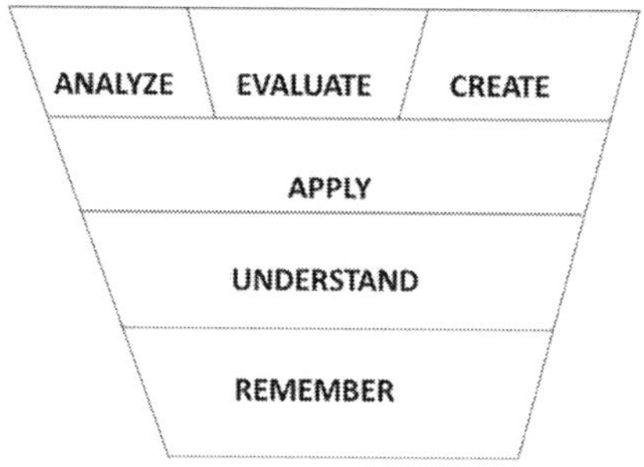

Figure 2.6 Bloom's Taxonomy Revamped.

members to learn new knowledge. After which, they go window shopping, select, evaluate, and finally decide to buy products. This behavior is almost the same, over domains. Consumers in areas like *domestic, healthcare, and education*, though have their specific cognitive behavior, whereby complaint and feedback mechanisms become vital components.

Instead of selection, they are only beneficiaries of a service provided to them and they can cause a change in the functioning of the system through feedback (education) and complaints (domestic & healthcare).

Conclusion and Discussion

CL is the involvement of the brain in the process of learning. Out of the nine cognitive skills a human has, six of the skills, namely, attend, listen, think, understand, solve, and respond, are included in CL. The brain uses a bootstrapping technique by which one "thinks what he/she is thinking", termed Cognitive understanding; which means that one is completely and consciously aware of what is going on in the mind. So, he/she is also capable of evaluating his thoughts and hence modulating them, in the right direction.

This is the norm of today, whereby one is not only judged by IQ (intelligence quotient) but also by one's Emotional Quotient. Akin to intelligence and the concerned question – "is intelligence existing from birth or is it acquired", CL skills can also be questioned whether it is natural or cultivated.

This chapter delves into proving that CL skills can be very well cultivated using different methods of teaching and presentation of information to students. Alternative techniques like Y&M, SKY, Sahaj, Sudarshan Chakra Kriya, and Mindfulness meditation, to name a few, were suggested to improve concentration and focus, and hence achieve efficiency in learning and understanding. Various learning theories have been summarized in the literature survey section, including the learning types in the introductory section. Impact evaluation and the steps to be followed for the same have been elucidated in the contributory section. Some of the applications have been mentioned for the sake of researchers, to pursue in a specific direction.

Bibliography

1. Victor, Stephen and Tim Spencer, "3 Cognitive Theories For Transforming Learning", *E-Learning Industry*, March 21, 2018.
2. Pavlov, Ivan, "Behavioural and Cognitive Theories", *Lifespan Development*: https://courses.lumenlearning.com/wmopen-lifespandevelopment/chapter/behavioral-and-Cognitive Theories
3. Skinner, B. F., *The Behavior of Organisms: An Experimental Analysis*. Appleton-Century, 1938.
4. Simatwa, Enose M. W., "Piaget's Theory of Intellectual Development and Its Implication for Instructional Management at Presecondary School Level", *Educational Research and Reviews*, Vol. 5, no. 7, pp. 366–371, July 2010, //www.academicjournals.org /ERR2 ISSN 1990-3839
5. https://www.simplypsychology.org/bruner.html
6. Zavershneva, Ekaterina and René van der Veer, "*Encyclopedia of Evolutionary Psychological Science*", Springer International Publishing AG, part of Springer Nature 2018, https://doi.org/10.1007/978-3-319-16999-6_2391-1
7. Loc, Nguyen Phu, Duong Huu Tong, and Vo Khuong Duy, "Using the Information Processing Theory into Teaching Mathematics: A Case Study Of —Vector Concept", *International Journal of Scientific & Technology Research*, Vol. 8, no. 09, pp. 1612–1616. September 2019, ISSN 2277-8616. https://www.ijstr.org/paper-references.php?ref=IJSTR-0919-22565
8. https://www.valamis.com/hub/cognitive-learning
9. https://www.indeed.com/career-advice/career-development/cognitive-learning
10. Driscoll, Marcy, *Psychology of Learning for Instruction* (3rd ed.). Boston, MA: Allyn & Bacon, 2005, pp. 341–373. https://ocw.metu.edu.tr/pluginfile.php/9013/mod_resource/content/1/driscoll-ch10%20(1).pdf.
11. www.wgu.edu/Five Educational Learning Theories
12. Paivio, Allan " Dual Coding Theory and Education", *Conference on Pathways to Literacy Achievement for High Poverty Children*, The University of Michigan School of Education, September 29-October 1, 2006.
13. Moreno, Roxana and Babette Park, "Cognitive Load Theory: Historical Development and Relation to Other Theories", https://doi.org/10.1017/CBO9780511844744.003, 2010.
14. Paas, Fred, and Paul Ayres, "Cognitive Load Theory: A Broader View on the Role of Memory in Learning and Education", *Educational Psychology Review*, Vol. 26, no. 2, pp. 191–195, 2014. http://www.jstor.org/stable/43549791
15. https://www.obsidianlearning.com/
16. Mayer, Richard E., *The Cambridge Handbook of Multimedia Learning*. Cambridge University Press, 2014.
17. Peters, Dorian, *Interface Design for Learning: Design Strategies for Learning Experiences*. New Riders, 2013.

Chapter 3

Professional Skills

A. Edward William Benjamin

Introduction

Throughout humanity, change and evolution have played a dominant role in the history of humans. In precedence with evolution, humans have formulated a new set of techniques to live their lives. They sculpted themselves to grow and to fit themselves dominantly in this ecosystem. After the infusion of technology, humans developed exponentially in terms of social and economic growth. At the initial stage of development, there was a need for Professional Education which played a significant role in maintaining human existence. There raised a need for doctors, Engineers, scientists, Teachers, Police, etc. For every essential need, a unique education system was formulated with its ethics. The skills that aid in professional development are professional. In every field like medicine, and engineering there exists a code of professionalism. These sets of codes or rules teach social and economic ethics which are to be followed by employees to uplift the society and to safeguard human life.

In this modern world, studies have shown that the opportunity to landa dream job varies directly with the quantity and quality of professional skills possessed by the employee. Even though one needs technical knowledge required for the job, professional skills play a key role in conveying matters and presenting themselves in a polished and professional manner so that the matter receives the necessary recognition.

Professional skills can also be defined as the employee's modest behavior, method of communication, simple yet elegant style of expressing ideas or thoughts, proper execution, and completion of tasks. Professional skills are the skills that can be learned and can be improved with more and more practice.

DOI: 10.1201/9781003504894-3

Professional skills are the skills that an employee can use to climb the ladder and think outside the box to provide creative solutions and to isolate oneself from the crowd. Technical skills and subject expertise are inevitable skills and traits that one must gain yet technical skills and subject expertise without presentable social and professional skills do not amount to the growth and development of an individual. One can always learn professional skills and gain the suitable traits that lead to growth and development.

Professional skills are divided into the following skills and can be used and utilized in various situations accordingly. They are as follows:

1. **Personal skills** which implies Initiative and Responsibility
2. **Interpersonal skills** which imply Leadership**,** Teamwork, and Conflict management
3. **Communication skills** which implies Written communication and Oral communication

Personal Skills

Personal skills are skills or traits that cannot be quantized. These are the skills that each individual possesses and these are the skills that reflect one's inner self and thoughts and actions. Personal skills show an individual's commitment towards a goal and actions and habits that lead to growth and development.

Apart from technical skills, personal skills play an inevitable part in one's professional career as they directly represent one's true nature, involvement, and interest in the field. Personal skills, their types, and some real-life scenarios are discussed below:

Personal skills are further split into Initiative and Responsibility which are the key factors that aid in the development of Professional skills in budding professionals. They are explained in detail below:

Initiative

Initiation symbolizes the beginning or start of an event that leads to the development of a subject or field or the process that ends an issue. Initiation is an essential skill that is missing in the current generation's young minds. They expect some other people to exhibit initiative or leadership skills.

Initiative as the name implies refers to the swift and essential action or move taken by the professional to solve or control an issue or problem that occurs which also includes prompt decision-making. Here are some examples of initiation

Some of the real-life scenarios in which the Initiative comes into effect are discussed below

- Let's imagine a situation in which there arises a need for a new idea or a problem that requires extra effort or creativity, an individual stepping forward to offer a creative solution to the existing problem and taking essential measures that are beyond his limits exhibits Initiative.
- Here is another example of initiation, when a problem occurs or one is confronted with an issue, swift action to solve that issue and preplanning that keeps one ready for any event of problem exhibits initiation.

Responsibility

Responsibility is a vital factor or skill that determines one's long-term existence in a field or a job. Responsibility is nothing but the amount of commitment one puts forward for a particular subject or task assigned. One must always share and acknowledge the failure and loss of a company as they do for the success and profits of a company. This trait shows the real nature of an individual and their involvement and dedication in the work they perform. Some of the areas that show this trait are listed below:

Examples where an individual exhibits responsibility are listed below:

- Here is an example of responsibility held by an individual 'X'. X plans his day-to-day activities and tasks to achieve his aim or goal on time without any delay so that neither time is wasted nor any essential resources. X accepts his flaws and mistakes at work.
- When an individual remains constantly motivated even at low times and does one's work perfectly, this is also an example of commitment and responsibility.
- Responsibility is also a major factor in groups that develops a group. When an individual is held responsible for the team and does one's part on time divides the workload participates in the team efforts and meetings and accepts the fame and also the downfall of the group. These acts show responsibility in a group.

Employers show interest in hiring employees who are accountable, dynamic, helpful, honest, loyal, punctual, reliable, responsible, teachable, and trustworthy. Though, being responsible might seem a small virtue it plays a tremendous role in establishing oneself and one's role in a professional working atmosphere.

Interpersonal Skills

Managing people is the key to organizational operation. Interpersonal skills are a vital part of this job. A person managing constantly interacts with others directly or indirectly. Interpersonal skills may be defined as interactive or social skills. They involve inspiring, motivating, leading, and managing others to achieve well-defined goals.

Certain jobs that involve dealing with people require interpersonal skills that may determine successful performance. Effective people who possess interpersonal skills win the confidence of others especially in matters of personal sharing, speak in meetings, listen actively, make sure that other persons take their views seriously when it matters to them, hold the attention of other members, avoid being either aggressive or passive, get others to confide in them, and keep in touch with others.

Effective Interpersonal Skills

Communicating with other people in an organization is as complicated as navigating a vehicle in traffic. It involves a greater understanding of the following guidelines. People have certain rights to be treated with respect, to be assertive, to be listened to, and to have information and perspectives.

Effective interpersonal skills include

- negotiations,
- listening skills,
- social skills,
- assertive skills, and
- multi-cultural communications.

Negotiation Skills

It helps us maintain interpersonal relationships which is crucial to organizational operations. Negotiation is also a kind of art where two parties come to a mutual agreement stating the pros and cons on each side. As effective negotiators, we should use our available preparation and planning time more productively than average negotiators.

Appropriate negotiation style gives us a sense of mastery over conflicting situations. It means analyzing available information, time, and power to influence others affect results, and arrive at a win-win situation. A pattern or structure in an orderly manner is essential to succeed in negotiation skills. This pattern includes the basic ground research work, Proper discussion and understanding of the goals, proceeding towards mutual benefits, and finally implementation.

Emphatic Listening

Another important skill that enhances interpersonal skills is empathetic listening. Whenever others are in a crisis or critical situation, those who are in a position to manage others should exercise active listening. Hearing is not listening. It requires a complete focus on the other person. An empathetic listener consists of the following communication aids:

1. *Patient listening*: It is paradoxically the stepping stone to active listening. The person who is involved in the process of active listening needs to focus the attention on the other person using the following components.
   a. **Eye Content:** Establishing good eye contact does not mean staring at others. Instead, it is looking naturally at the other throughout the conversation.
   b. **Attentive Posture:** while listening, it is good to slightly lean towards the speaker. Facial expressions should indicate that the listener is fully attentive and puts the focus on the other person.
   c. **Supportive Gestures**: Body language speaks more effectively than the actual words that are used. The listener has to show a great deal of interest by gentle gestures.
   d. **Appropriate setting:** The place and time chosen for the process of empathetic listening should be supportive. The place should be conducive for sharing.
   e. **Meaningful Silence:** Silence is also communicative. Attentive silence can send a message that the listener is fully with the speaker and give her or him time to think and proceed with the presentation comfortably. Meaningful silence is particularly useful in situations of grief and loss.
2. *Active Listening*: Those attending skills form only the first part of the Active Listening process. There are a few other important responding skills that complete the circle of this process. It demands a total commitment from the listener to the communication. It, moreover, requires the listener to view the world as perceived by the other person and communicate this understanding to her/him. Active Listening Communicates the Listener's acceptance of the person too. Given below are responses that help persons build bridges.
   a. **Acknowledgment response**: these are brief, one or two-word statements and non-verbal gestures to show the other that the listener is following the conversation. 'Oh', 'Okay', 'Really', 'Yes', 'Sounds nice', etc., are some examples of this kind of response.
   b. **Door opener:** It is another responding skill and a pleasant invitation to others to talk. Instead of asking questions, the listener can encourage the other to open the discussion as given below: 'You look excited', 'It looks as if you are confused', etc.
   c. **Reflecting content:** it means listening accurately to the other and reflecting the essence of the content of the thoughts, ideas, beliefs, and facts of the other. It enables the listener to ensure her or his understanding of the other's communication and not to be mentally absent while listening.
   d. **Recognizing feelings:** It is another important aspect of Active Listening. Individuals have the right to feel and express their feelings. Listening

actively to such persons will enormously help them release their bottled-up emotions. Recognizing and reflecting accurately the negative feelings such as hate, disappointment, fury, etc., will heal the other's strong feelings, and recognizing and reflecting the positive feelings such as love, happiness, appreciation, etc., will deepen their joy.

e. **Summarizing:** It is done after reflecting and recognizing the content and the feelings of the other. It aims at paraphrasing the total meaning of the communication of the other. It is saying in one's own words, how she or he has captured the other's thoughts and feelings about a particular situation or a crisis. Paraphrasing is not repeating the other's thoughts word by word, rather it is an attempt to communicate the understanding of what the other says and how she or he feels about it.
f. **Lending a hand**: It is another important felt need in the process of helping others in crisis. Recognizing and summarizing the total meaning of the other's communication alone does not satisfy a person, especially while experiencing a crisis all by herself or himself, the listener can respect their genuine need and lend her or him a hand.

Social Skills

Social skills are instrumental in initiating, building, maintaining, and enhancing our interpersonal relationships in the workplace. There are many levels of relationships that operate in work situations. Some may border on friendships and some may work on partnerships. Our ability to maintain the relationship depends on personal skills. However, some requirements come from our professional relationship with each other. They include

- Informing: telling others what they need to know.
- Consulting: seeking opinions from those who are interested in the issue.
- Agreeing: arriving at consensus in decision-making.
- Eliciting: getting others' perspectives.
- Respecting: despite differences, accept others.

Above all, we need to be sociable. It is more a question of taking time and interest in others, showing willingness to spend time with them, and an interest in their need

Assertive Skills

When we deal with other people, different types of behavior are usually adopted. Aggressive behavior indicates people standing up for their rights at the expense of those of other people. Passive behavior emanates from people who believe that their rights are less important than those of others. Assertive behavior shows standing up

Table 3.1 Difference among Aggressive, Passive, and Assertive Skills

Aggressive	*Passive*	*Assertive*
Boastful	Apologizing	Problem-oriented
Excess use of 'I' statements	Few 'I' statements	Clear statements and to the point
Verbal threats	Long, rambling statements	Offering suggestions
Blaming others	Dismissing one's own needs	Constructive criticism
Sarcastic		Plain
Shouting	Frequent justifications	Attempt to find out others' needs
Threatening gestures questions	Self-putdowns	No' should' or 'ought'

for one's rights and needs while respecting those of others. Hence, we should emulate assertive behavior that depends on an understanding of the rights of all people.

We can recognize the distinguishing aspects of these behaviors and try to follow the acceptable behavior for maintaining effective interpersonal skills (Table 3.1).

Assertive behavior has the advantage of being influential because it is consistent while being respectful.

Here are some simple techniques and methods for developing assertive behavior:

- Know the facts and handle them appropriately.
- Anticipate other people's behavior and prepare your response.
- Use open-ended questions that would invite opinions from others.
- Condition your mind so that you don't react to aggressive behavior.
- Develop self-confidence that will work well to bring forth results.
- Learn to respect the space for others.
- Read inspirational stories or information about performing leaders that will reinforce your faith in proper values and your natural style and self. For example, recall information about the captain of the Indian Cricket Team. M.S Dhoni, who unpretentiously but assertively led India to World Cup 2011 victory.

Multi-cultural Communication

People–working in organizations these days are drawn from culturally diverse backgrounds and therefore, they need to understand the means of relating and communicating well. Multi-cultural collaboration is important because the nature of work that has spread over the world involves networking and sharing of resources. In this modern world of globalization, each individual ought to respect and behave each other's culture and values.

- It treats everyone equally.
- It makes for more effective communication among groups.
- It creates community.
- It leads to a more just society.

Enhance Your Interpersonal Skills

Interpersonal skills are of paramount importance in the workplace. They are important to both people who are in a position to manage others as well as the persons who work under the guidance of such supervisors. Any amount of hard work and brilliance without interpersonal skills may not put you up on the runs of the ladder of success. You may improve the necessary social skills and become a professional by following simple things such as:

- Maintain a cheerful attitude.
- Be appreciative.
- Pay attention to the views of others.
- Bring people together.
- Resolve conflicts
- Communicate clearly
- Have a sense of humor with colleagues
- Avoid finding fault with others
- Recognize the genuine needs of others

Communication

Since human evolution, man has invented and practiced various methods and techniques to effectively communicate with one another. This process of implementing innovative communication has led to the exponential growth of the modern world. Communication is the art of conveying one's thoughts to others in a rather simple way such that the person who receives the message understands it without any confusion. Effective communication plays a key role in achieving the desired result in a particular matter.

- It is vital to all functions.
- It is a continuous process.
- It is to create mutual understanding.
- It is a two-way process with the sender and the receiver.

The effective methods of communication are discussed below

Communication Skill

As depicted in the picture above, effective communication flows properly. The sender or the person who wants to convey a thought or idea prepares it in a rather simple way so that the receiver understands it without much difficulty. The message is transmitted via an intermediate which may be electronic forms like email, messages, or trivial ways like letters or even oral communication. Then the receiver receives the information processes the facts and acts upon it. The various ways or methods in which effective communication proceeds are given below (Figure 3.1):

Communication Networks

Communication is the pattern through which the members of a group communicate.

The Centralized Pattern

In this pattern, all the messages propagate from a single individual who might be the head i.e. it is centralized (Figure 3.2).

The Y –Pattern

In the T-pattern, messages are passed from two individuals and proceed in an orderly manner (Figure 3.3).

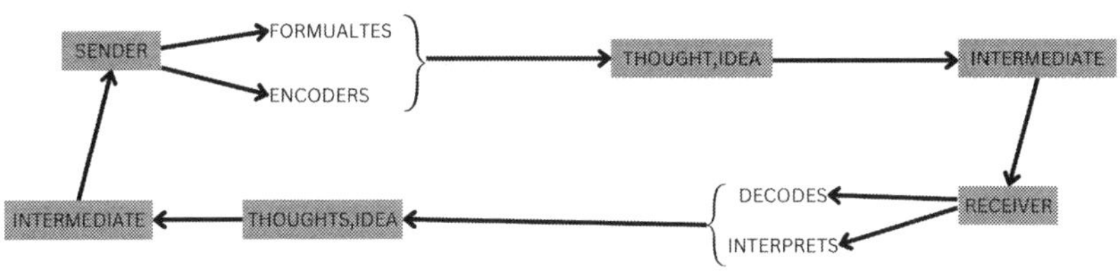

Figure 3.1 Schematic representation of communication skill.

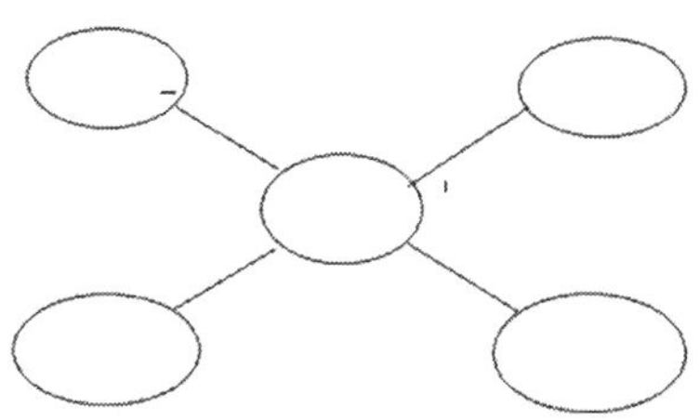

Figure 3.2 Diagram illustrating a centralized pattern of communication network.

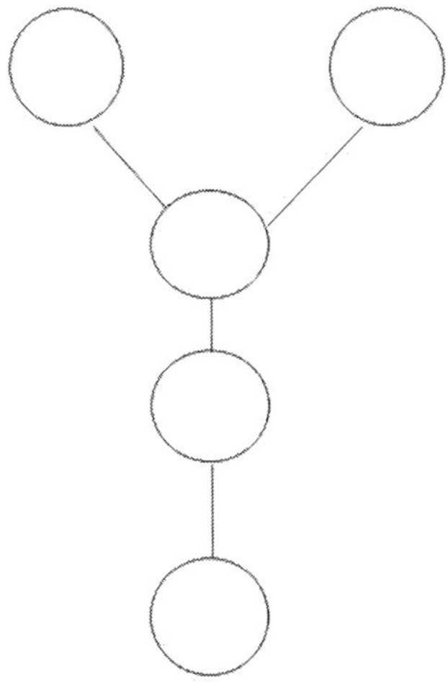

Figure 3.3 Diagram illustrating y- y-pattern of the communication network.

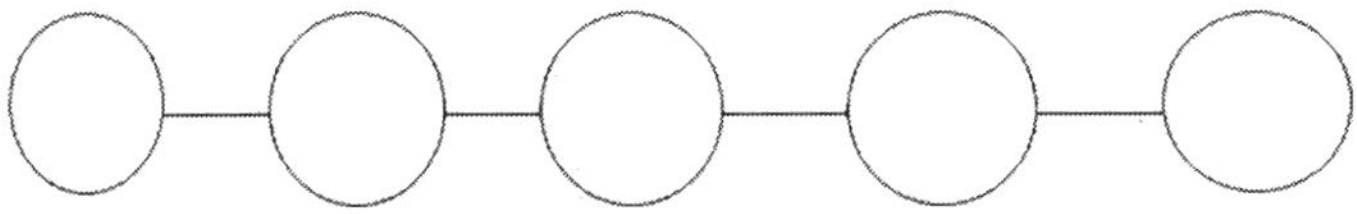

Figure 3.4 Diagram illustrating a straight pattern of the communication network.

The Straight Pattern

This straight pattern provides an orderly sequential pattern of flow of information which is highly essential for the proper functioning of a system (Figure 3.4).

The Circle Pattern

This is similar to the straight but the pattern is closed. This pattern results in brainstorming and innovation in the subject. It is the most decentralized. Everyone participates equally (Figure 3.5).

Some Other Varieties of Effective Communication

One-Ways

Communication becomes active and the receiver remains passive. This can be illustrated as given below (Figure 3.6).

The teacher must make it a two-way process.

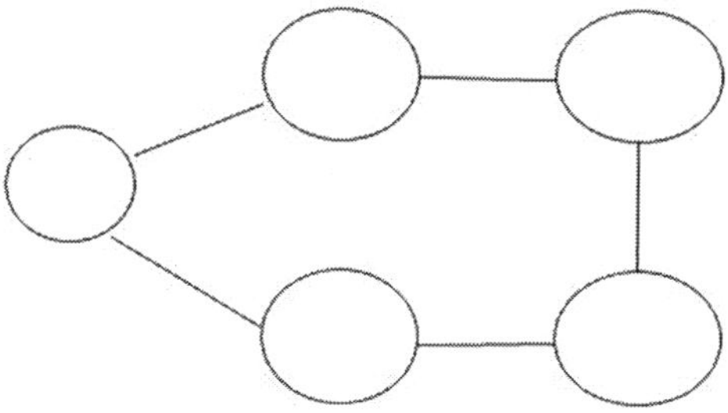

Figure 3.5 Diagram illustrating the circle pattern of the communication network.

Figure 3.6 Block diagram of One-way of effective communication.

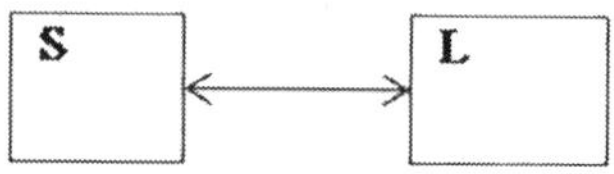

Figure 3.7 Block diagram of discussion of effective communication.

Discussion

It is a two-way process in which there is constant interaction between the teacher and the pupils. This can be represented as follows (Figure 3.7).

Group Procedure

In some group procedures like, seminars, symposiums, workshops, panel discussions, and debates, communication takes place between the employees. This may be represented as follows (Figure 3.8).

Problems and Projects

In this method, both the employers and the employees Co-operate with one another. This can be diagrammatically given as follows (Figure 3.9).

Communication Barriers

Despite the innovative and effective methods of communication, there always seems to be a problem. Perfect and smooth communication is inevitable for growth yet

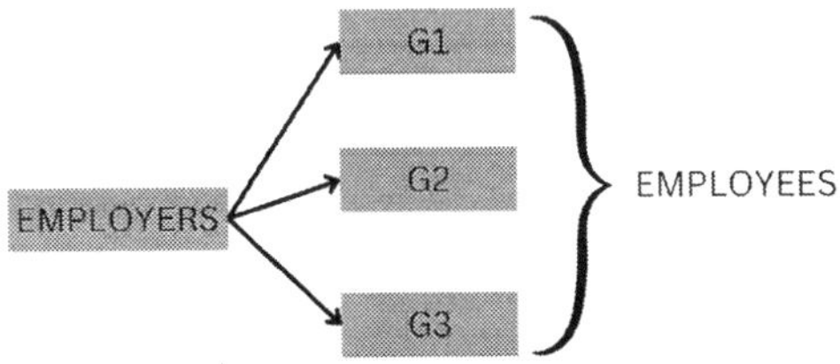

Figure 3.8 Block diagram of group procedure of effective communication.

Figure 3.9 Block diagram of problems and projects of effective communication.

it is difficult to achieve it at times. This may happen owing to conflicting subjects, improper attention to improper attention due to reluctance in poor listening traits, and some psychological factors. These can be overcome by some methods that are given below:

Measures to Overcome Barriers (Principles of Effective Communication)

The following steps may be taken to overcome the barriers of communication.

Clarity: the message must be clear. Language should be as simple and precise as possible.

Completeness: The message must be comprehensive and adequate. In complete message creates misunderstanding and delays action.

Brevity: all communication should be brief. Meaningless words and an overload of information spoil a message.

Timeliness: The message must reach the receiver at the right time.

Compassion: the communicator must understand the intelligence level and the background of the receiver.

Feedback: communication is a two-way process. There should be follow-up action to ensure that the message is rightly understood.

Attention: careful listening is essential for effective communication. The communicator should try to secure the undivided attention of the communicator.

Use of Grapevine: A formal channel of communication should be supplemented by a grapevine (informal channel). It takes place in informal and inter-personal contact

Conclusion

Therefore, Professional skills are the skills that should be given equal importance as technical skills. Technical skills are the content whereas professional skills are how we present the content. From this, we can understand that professional skills also play a significant role in the development of an employee. Excellent personal skills, Proper interpersonal skills, and effective communication skills are the three major elements in establishing an individual's talent and proving one's self. These are the skills that can be mastered over time and thus can be utilized for the all-round development of an individual.

Chapter 4

Fostering Critical Thinking Skills in Modern Learners

A Peep into Higher Educational Institutions

Happy Saikia and Nil Ratan Roy

The Backdrop of the Study

> The function of education is to teach one to think intensively and to think critically intelligence plus character-that is the goal of true education.
>
> – *Martin Luther King*

To think critically requires experience as well as logic and creativity. Critical thinking is the key aspect of discovering resolutions to solve issues and challenges. The power of thinking critically is one of the most vital skills for 21st-century young learners. In this rapidly changing world, the power and the ability to think critically is a core life skill one can have; the young learners to step into the diverse field of their life, to reach their goals; it is necessary to hold the power of thinking creatively and flexibly to adopt in the given situation (Sigit, 2019).

Higher educational institutions should take initiative for the modern learners to sustain skills to ensure shaping individuality for making any decisions or resolving issues. For modern educational approaches cultivating thinking skills in young modern learners is an important element. Critical thinking is the key to successfully expressing individuality while making a decision and solving a problem. Higher educational institutions should take the initiative to foster critical and creative

 DOI: 10.1201/9781003504894-4

thinking skills by providing them with opportunities to make decisions to solve their real-life issues (Hirose, 1992).

Critical thinking boosts creativity and enhances the way of managing time, thinking critically does not represent the ability to think logically rather it represents the power of applying these skills to real-life circumstances which are not self-reliant (Lisa & Mark, 2019).

Critical Thinking Skills as a Core Life Skill

> Life skills are abilities for adaptive and positive behavior, that enable individuals to deal effectively with the demands and challenges of everyday life.
>
> *(WHO, 1994)*

Teaching life skills as generic skills about everyday life could form the foundation of life skills education for the promotion of mental well-being, and healthy interaction and behavior. More problem-specific skills, such as assertively dealing with peer pressures to use drugs, or to become involved in vandalism, could be built on this foundation. There are research indications that teaching skills.

Life skills are the core elements or skills for promoting mental well-being and shaping a sound personality; research proves that instilling life skills in young modern learners in the 21st century as part of education is an effective approach. Higher educational institutions can take the initiative to instill the core skills in the adults of nations to build a healthy society (UNICEF, 2020).

The model below shows the place of life skills as a link between motivating factors of knowledge, attitudes, values, and positive health behaviors; in this way contributing to the primary prevention of issues in learners:

The model below shows the way of prevention issues of learners through life skills as a link between positive health, values, attitudes, and knowledge (WHO, 1994; Figure 4.1).

Objectives of the Study

The present study has been designed to carry out the following objectives:

1. To explore the key factors and instructional practices that are needed in fostering critical thinking skills among modern learners in higher educational institutions.

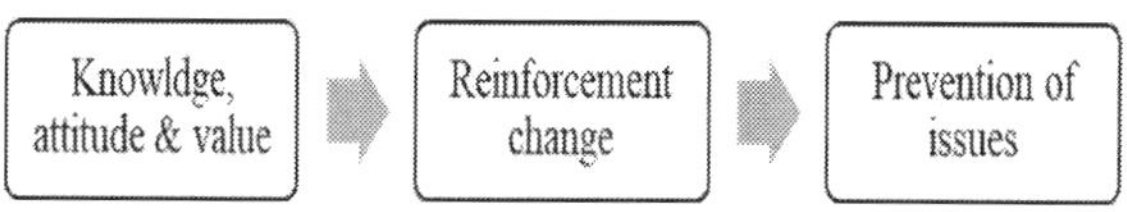

Figure 4.1 Prevention issues.

2. To analyze how critical thinking skills assist modern learners in the decision-making process.
3. To study the challenges faced by higher educational institutions in developing critical thinking skills among modern learners.

Methodology

The present study has been carried out by exploring data from secondary sources like Journals, and articles related to critical thinking skills and modern learners. To find out the key aspects of fostering critical thinking skills among modern learners in higher educational institutions; attention has been given to the comprehensive literature related to the study.

Section 1: Fostering Critical Thinking Skills: Instructional Practices and Key Factors

One of the most significant goals of higher educational institutions is to cultivate skills in learners to understand important ideas, analyze and think critically, and use and be powerful with all the learning skills to grow holistically in life and to be a great contributor to the culture and society.

Twenty-first-century learners are approached to learn by experiences, when the learners think critically; they are encouraged to think for themselves, to analyze, and to incorporate; questioning is the main path of critical thinking that offers the source of knowledge formation; the main factors that influences higher educational institutions is to create a road map for the learners to be taught and to be learned through any given situations or issues. Research and projective work carried out in higher educational institutions lead the learners to develop critical thinking and creativity in them (Figure 4.2).

As Elder and Paul say, modern learners are not inactive in possessing thinking about originality; the curriculum is thus a minding alerting device. Which realizes to pay attention to learners' individuality and interest to determine the outlines of the different aspiring minds and lives. The curriculum provided in most of the higher educational institutions prepares learners to be independent rather than being a shadow of pre-determined roles. Assisting the learners to be independent and letting them possess their thinking skills through the maze of issues and challenges to present individuality in life (Paul & Edler, 2018).

Higher educational institutions should take the initiative for the young modern learners to make them prepare to meet every change of life rather than taking their skills and knowledge for granted. Higher educational institutions have the major role of teaching thinking critically to the learners since students go to broader fields

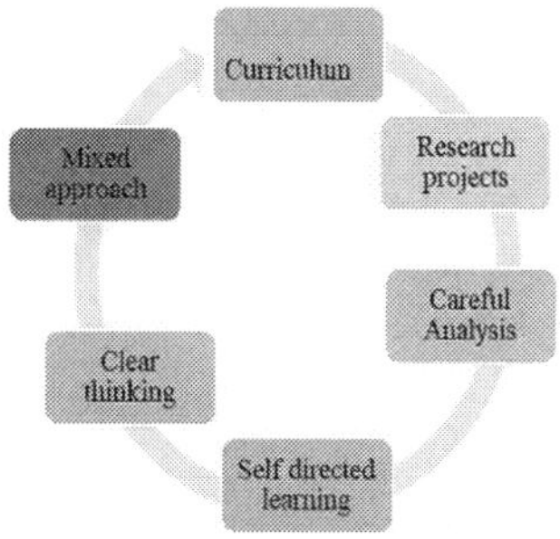

Figure 4.2 The key factors influencing critical thinking in Higher educational institutions.

after institutions to meet every challenge of life and solve societal issues; by allowing them to organize events in their way letting them learn through the maze of challenges and institutions can have a course unit separately or having a mixed approach can inculcate skills in them as an integral part of the curriculum (Hove, 2011).

Section 2: Critical Thinking Skills and Decision Making: Role of Higher Educational Institutions

One of the most powerful elements of human life is the capability to make a healthy decision; which requires one to possess critical and creative thinking skills to overcome glitches. To come at a decision takes place impulsively or suddenly; since it has a greater role in life it is obvious that to make a better decision requires thinking better to have an advantage significantly.

Higher educational institutions instill skills in the learners to face the outer world, to be adaptable, and to be a good leader the core skill requires knowing how to think and what to think, to know how to think is an art; the use of decision-making, administrative, and convincing skills can assist to instill leadership quality in modern learners. Developing decision-making or problem-solving requires cooperation with the people around them, therefore, practicing brainstorming sessions, and developing communication skills and leadership skills assists the learners in being good decision-makers.

Teachers and mentors' observations of students' activities in the implementation of any decision taken by them; examine the pros and cons of the result, assist in overcoming the drawbacks, and analyze logically improve students thinking process by reviewing the techniques of the decision-making process.

While making a decision should always focus on rationality rather than being emotional or intuitive, Critical thinking is the core element for decision-making. Making an effective decision requires knowledge, experience, and information that is logical or rational. Allowing students to organize events in the institutions, involved in projective work led the students to think critically.

In the 21st century providing students in such situations to have more discussions and brainstorming sessions helps to think logically and think 'out of the box' to bring changes in the conventional ways to have a better future. Years ago, when higher educational institutions were not well equipped with modern technology, the learners and the philosophers through their critical thinking process established the modern world to get access to almost everything.

Higher educational institutions have the main role in producing great thinkers to remove all the gaps in society by allowing and motivating the learners to engage in activities like research work, curriculum development, carrying out projects, etc., which requires a high critical thinking process to make decisions to solve issues.

Section 3: Challenges and Its Prospect to Instill Critical Thinking in Modern Learners

The curriculum that is followed in our educational institutions needs to be developed extensively to enable young learners to possess thinking skills. The curriculum that is followed in the institutions requires modification to ensure learners are well-adaptive in any given situation after successful completion of higher education.

Another major cause that hampers in development of critical thinking skills in modern learners is dependency on technology. Higher educational institutions are dependent on technology in every aspect of work they involve; the spoon-feeding technology sometimes affects in critical thinking and creative development of the learners.

Higher educational institutions can take the initiative to develop the skills they need in life by providing innovative and modern techniques such as the fishbone technique which allows them to know the root causes to solve a complex issue, it has a greater impact on students to study systemically each issue and challenges faced in life. Research proves that by adopting such techniques critical thinking skills can be developed in modern learners.

The research report shows that brainstorming is an innovative technique of thinking that helps articulate ideas to come to decisions. In the classroom, the learners must be allowed to practice such innovative techniques while making a decision or in other related activities, which helps to have the power of critical and creative thinking by listening to various ideas.

Concluding Remark

Fostering critical thinking skills in higher educational institutions is not only for setting a criterion for scoring well in academics or related works rather it is a necessary aspect of holistic development. In the last decades, critical thinking skill has been given the utmost importance to adjust modern learners to this technical and complex 21st-century global world.

References

Hirose, S. (1992) Critical Thinking in Community Collages. ERIC Digest. Retrieved from http://www.ericdigest.org/19922/critical.htm

Hove, G. M. (2011) Developing Critical Thinking Skills in the High School English Classroom (Master's thesis, The Graduate School University of Wisconsin-Stout Menominee). Retrieved from https://www2.uwstout.edu/content/lib/thesis/2011/2011hoveg.pdf

Lisa, G. S., & Mark J. S. (2019) Teaching Critical Thinking and Problem-Solving Skills. Retrieved from https://dme.childrenshospital.org/wp-content/uploads/2019/02/Optional-_Teaching-Critical-Thinking-and-Problem-Solving-Skills.pdf

Paul, R., & Elder, L. (2008). *Critical Thinking: The Nuts and Bolts of Education. Optometric Education*, 33, 88–91.

Sigit, P. (2019) Students Critical Thinking in Writing Background of Research, *Education and Humanities Research.* http://doi.org/10.2991/assehr.k.200427.061

UNICEF. (2020) Comprehensive Life Skills Framework. Retrieved https://www.unicef.org/india/sites/unicef.org.india/files/2020-02/Comprehensive-lifeskills-framework.pdf

World Health Organization. Division of Mental Health. (1994) Life Skills Education for Children and Adolescents in Schools. Pt. 1, Introduction to Life Skills for Psychosocial Competence. Pt. 2, Guidelines to Facilitate the Development and Implementation of Life Skills Programs, 2nd rev. World Health Organization. Retrieved from https://apps.who.int/iris/handle/10665/63552

Chapter 5

Creating Interest in Students – A Gaze

L. Vidyarani and N. Mohanapriya

Introduction

Creating interest among students in education leads to academic success. A powerful tool for winning in a career is interest. Inculcating interest should be the first process of teaching and learning. Interest in any activities makes changes both mentally and physically. Interest in any academic activities will help in reforming societal needs. A person with lots of interest and who excels in academics will have the potential to tackle any complex situations in their life.

Instilling interest within the students is a challenging task for the teachers nowadays, since they may suffer from physical and mental problems. This article describes various tasks that have to be undergone by the teachers to take the initiative to drill the young minds and convert them into Nation-builders.

Aesthetic Atmosphere

Institution

The aesthetic quality of education helps the student to create a different type of perception in all activities and makes them create new innovative ideas. Aesthetic environment enhances the following qualities of education in the institution (Chou, 2016).

 DOI: 10.1201/9781003504894-5

- Power of ability to work by students and teachers
- Student and teacher sharing relationship
- Taking Open And Sincere Communication
- New ways of learning with more excitement.
- Self-confidence gained from day-to-day experiences
- Analysis of human correlations
- Opportunity to combine mind and emotion, cognition and sensory experience, analysis and intuition toward understanding something as a whole
- Develop a reading atmosphere
- Encourage different types of learning styles.

Home

Aesthetic environment really changes the feel of a student and greatly impacts the attitude and mindset of learning. As Children are susceptible to their surroundings, an aesthetic feel at home can make them feel good, and calm and enhance their interest in learning (Apps, 2012). An aesthetic feel can motivate and encourage a student with a positive attitude and will always fascinate and produce excitement in learning.

Mentor with Empathy

Facilitator

A mentor should be a facilitator, which enables the mentee to create interest in academics and to make proper decisions with clear thinking by receiving proper guidance and achieving a solution to a problem with ease (Ghosh, 2018).

The facilitator will help the students to develop a vision and lead the group toward achieving it. A facilitator helps a group of people to work together better, and understand their common objectives. When a mentor is a facilitator, students feel good to share their thoughts and convey their ideas in implementing it.

Honest and Candid

An honest mentor is a very positive quality of mentorship. Being honest will help the mentee to heal their pains and share their problems. Honest mentor should be open-hearted with lots of empathy which enable the students to relieve their mental stress and get them engaged in their academics. A candid teacher treats the students in an unbiased manner and helps them to speak their minds and their actual opinions.

Rejuvenator

A mentor as a rejuvenator can reset young minds to explore their talents and brighten their future. Rejuvenators will always be in accord with students and support them by being their spark and stimulating their skills to improve their societal behavior (Young, 2017). When the mentee gets rejuvenated they learn how to contribute positively to communities.

An educator encourages healthy habits and exposes students to new things, learning from diversity, making things fun, and insisting on the importance of family and support networks.

Learn While Play

Improves Cognitive Development

Students gain and explore their knowledge through their interest in learning while playing. This helps to find the solution to a complex problem in real-life situations. When children learn while they play he/she may understand and think about the happenings around them.

Learning while playing will enhance the cognitive skills in students which leads to the progression in learning skills with lots of paying attention to studies and memory power. Students will gain the quality of learning to evaluate, analyze, compare, and understand through cognitive skills. Students with cognitive skills can improve their ability to focus on things and complete the task on time.

Develops Communication Skill

Learning while playing will help students improve effective communication in students. Cooperative play enables the students to enhance the quality of communication with their peers. The playway method helps the students to improve their vocabulary and supports to upgrade their ability to communicate, express, and understand feelings.

Way method in education inculcates various soft skills in a student and makes them get connected with peers and teachers easily. It facilitates overall and holistic development.

Enhance Relationship Building

Learn while play is an activity-based method of learning which gives hands-on experience for the students. Students mingle with others communicate with their surroundings and learn from various members including student friends and teachers which in turn creates a good bond within them.

A positive learning environment is created when the students build good relationships within the classroom. Building relationships with students is a tedious task and can be made possible by bringing learn while play methodology to make the task simple for the mentor.

Experimental Learning

Do and Feel

The influence of experiential learning eradicates rote learning and intensifies the knowledge and skill blended with feelings and emotions. It helps the student to apply the knowledge practically with a high level of retention. It also creates team collaboration and communication skills among the students which enable them to lead their lives hand in hand with society.

Experimental learning plays a vital role in retaining concepts and ideas and bridges the gap between theory and practice (Kalyani & Rajasekaran, 2018). Experiential Learning provides a demonstrable mindset within the students with a high level of outcome. Experiential learning methodology is highly effective in the journey of Teaching, Learning, and assessment to enable personalized learning and to set their own learning pace.

Liberal Education

Free Entry with Open-Mindedness

Open-mindedness helps the students to create opportunities and reform their thoughts to find a better solution to a problem (Thompson, 2017). A student with open-mindedness will gain more general information and enhance the soft skills which help in analyses of different approaches, methods, and strategies.

Students who are given freedom can expose all their ideas and will be able to tackle vulnerable situations. Open-mindedness increases the learning interest of students with a cooperative approach combining shared ideas which enhances their leadership quality. Open-mindedness shows rapid growth when students mingle and learn collaboratively with other community students.

Reflection of Thoughts

Reflection of thoughts is a learning that comes from experience and helps the students to reach leaps and bounds. It helps to create new innovative ideas from experiences. This enhances the quality of framing questions within us which in turn helps to create new ideas and bring them into a design.

Having group discussions among students makes them expose their thoughts which is the most significant Quality for personality development. Through this practice, students will always be active and get engaged in all activities. It never leaves a student's mind idle. Reflection of thoughts gives progress with a strong foundation in the teaching-learning process with updation.

Exposure of Ideas

Brainstorming is one of the unique techniques to expose the different ideas of the students (Kalyani & Rajasekaran, 2018). It helps to create new innovative ideas and get the capability of differentiating the good and bad. Sharing and exposing ideas build confidence. Giving freedom to the students to share their ideas will uplift their innovative thinking.

Enhance Erudition

Create Curiosity

As a facilitator, we must create curiosity about learning among the students. The classroom environment should be converted into a place that explores joyful learning which leads to their well-being in life (Learning, 2010). If all the students thinking and contributions are accepted in the classroom, the curiosity of the student's learning gets increased.

When theoretical teaching is converted into experimental learning, curiosity in students for learning is enhanced. Curiosity to learn doesn't end up only with academics; it helps in upliftment throughout his life. Curiosity can be developed within the students by eradicating their fear and converting boring topics into interesting stuff.

Cognitive Skill

One of the basic qualities to enhance the skill of a student is cognitive skill. Cognitive skills can be developed through various physical and mental exercises (CBSE, 2020). It helps the students to understand the concepts and prototyping of their ideas. Cognitive skill helps the mentees to develop their capabilities and performance in the classroom and make them shine in society. Cognitive skill helps students for the development of analytical skill, capacity building, managerial skill, Leadership quality, and personality development. Cognitive skill is the strength of the student.

Learn through Travel

Learning is a lifelong process. The behavioral skill of a student gets improvised through travel. Dealing with other environments and peers gives new learning with lots of creativity and imagination. Learning beyond the classroom energizes the brain and makes the brain active with more enthusiasm and thirst to learn. It helps the students get mingled with other communities which enhances collaborative learning with a deeper understanding of concepts in cross/ interdisciplinary areas. Sharing of knowledge with good understanding and maintaining the relationship with humanity gets enlarged.

Learning through travel is different from the conventional teaching method which makes the students flourish with a wide range of skills such as Leadership quality, humanity, teamwork, etc.

Differential Learning Pattern

Collaboration

Collaborative learning makes the students work together with quality interactions. Interpersonal communication in team-based learning helps to develop innovative skill and co-learning skills. Collaborative learning creates more interest in sharing knowledge with peers (Arrington, 2015). When there is knowledge sharing among the students, ideas get explored and create curiosity towards learning. It reduces the fear and helps in exposing the skills. Collaborative learning helps to enhance *problem-solving and critical thinking*. Students undergoing collaborative learning will always think of being together and building unity in society.

Innovative Skill

Innovation skills in a student comprise various skills like *creativity, critical thinking, communication, and Analytical skills which help them solve complex problems with correct solutions. It plays a vital role in the individual development and the surroundings. It helps in personality grooming and managerial skills in all the activities. Innovative skill improves the efficiency of the students and break through hurdles.*

Conclusion

The current Education system has to be enhanced with the above-mentioned skill sets to fulfill the needs of the students. Skill development programs and modifications in teaching with modern teaching methodologies in the education system will help the students to create interest in their learning process.

Students are the future Nation Builders they should be very well molded with knowledge, innovative skills, morals, and ethics with humanity. Students' skill sets inculcated in them must help them to be loyal and patient in all their activities.

References

Apps, Linda and Margaret MacDonald, "Classroom Aesthetics in Early Childhood Education", *Journal of Education and Learning* Vol. 1, No. 1, 2012. http://doi.org/10.5539/jel.v1n1p49.

Arrington, Nancy McBride and Adrienne Cohen, "Enhancing the Scholarship of Teaching and Learning Through Micro-Level Collaboration Across Two Disciplines", *International Journal of Teaching and Learning in Higher Education* Vol. 27, No. 2, 194–203, 2015.

Chou, Mei-Ju; Jui-Ching Cheng, and Ya-Wen Cheng, "Operating Classroom Aesthetic Reading Environment to Raise Children's Reading Motivation" *Universal Journal of Educational Research* Vol. 4, No. 1, 81–97, 2016, http://doi.org/10.13189/ujer.2016.040111.

Ghosh, Rajashi, "Diversified Mentoring Relationships: Contested Space for Mutual Learning?" *Human Resource Development International* Vol. 21, No. 3, 159–162, 2018, https://doi.org/10.1080/13678868.2018.1465670.

http://cbseacademic.nic.in/web_material/Manuals/21st_Century_Skill_Handbook.pdf

http://www.eurosfaire.prd.fr/7pc/doc/1300702480_jrc62370_learning_teaching_2010.pdf

Kalyani, D. and K. Rajasekaran, "Innovative Teaching and Learning", *Journal of Applied and Advanced Research* Vol. 3 (Suppl. 1) S23–S25, 2018, https://doi.org/10.21839/jaar.2018.v3S1.162.

Thompson, Winston C. "Liberalism in Education", http://doi.org/10.1093/acrefore/9780190264093.013.49.

Young, Cheri A.; Badiah Haffejee, and David L.Corsun, "Developing Cultural Intelligence and Empathy through Diversified Mentoring relationships", *Journal of Management Education*, 1–28, 2017, http://doi.org/10.1177/1052562917710687

Chapter 6

Role of Need Analysis in Invoking Interest of Students to Learn English

K. Kamala Devi and A.R. Saravana Kumar

Introduction

Interest is a supreme factor that propels the process of learning, directs both academic and professional enhancement, and finally an indispensable component of academic success. Without inculcating interest, it is very difficult to keep the students focused on the program, and highly challenging for the teachers to successfully transact the curriculum. To provide a fruitful learning experience and keep the learners self-motivated and deeply involved throughout the learning process, the entire education system should focus on the interests of the students. The primary drive of the interest is the 'need' of the students. Hence, it is quintessential to do a need analysis of the learners for effective instructional design. Needs assessment is one of the main investigative tools used by teachers or educational administrators/institutions for the identification of actual needs, gaps, and hidden parts of the system and other activities. From this investigation, curriculum developers obtain various pieces of information on linguistics background, learning style preferences, teaching strategies, students' motivations, and students' commitment to their involvement in learning. Needs assessment will help teachers analyze the learning materials, content, and media for their learning programs, and in turn, will be instrumental in developing suitable learning strategies.

DOI: 10.1201/9781003504894-6

Importance of Needs Assessment

The methodical process of procuring information that is appropriate and adequate to design an effective educational program that will address the learners' needs and wants (gap). The methods which are frequently used for determining a needs assessment include the following

1. Conduct of Needs Survey
2. Analysis of the requirements
3. Request from the Potential audience (focus groups)
4. Changes in policy
5. Review of relevant literature
6. Reflection on the previous learning activity
7. Current trends of the learner characteristics
8. Novel methods of learning, teaching, and evaluation

The significance of Need assessment lies in identifying the gap between the expectations of the learners and the entire instructional design including the objectives, instructional transaction, and assessment procedures. In "A Guide to Performing a Needs Assessment and a Gap Analysis" Anthony J. Jannetti expressed that there can be gaps in either knowledge, practices, or skills. For efficient progress towards achieving the academic goals and providing a successful learning experience, it has to be assessed in terms of what is working well and what needs to be modified. A Gap analysis may be defined as determining the difference between the present knowledge (what is being done) or practices and current evidence-based practices (what should be done; Figure 6.1).

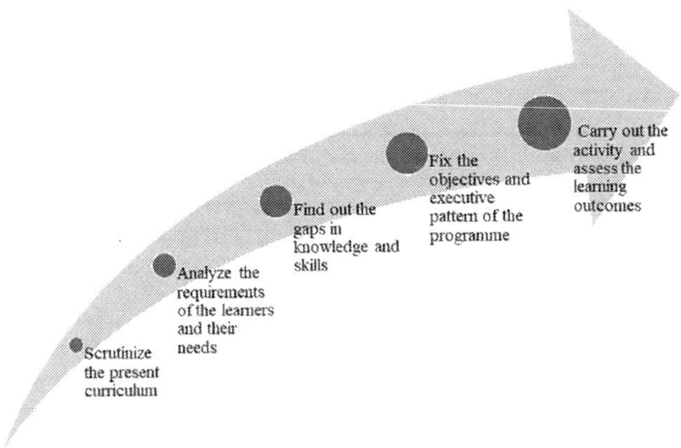

Figure 6.1 Phases of finding out the gap.

Changing Needs of Millennials

During the 20th century, students learned English as a second language. They are varied in their abilities and interests and most cases, English was taught as one of the subjects with the domination of the regional languages in English classes. The focus was on the improvement of reading and writing skills as the professional education students needed to read the study materials and write the examinations in English. Neither the teacher nor the teacher worried about the need to practice English.

Even in the 21st century, the challenges from the previous century continue. At the same time, a new learner has also emerged: the Indian millennial whose access to tech and connectivity has increased along with his/her expectations from online products and offline services. The expectations of the millennials of India are (i) learning should be an experience and not just knowing, (ii) to work/learn in groups collaboratively, (iii) immediate communication with the teachers and getting feedback for their performance as early as possible and (iv) abstinence from routine methods of learning. Frand's list of beliefs of millennials is enormously true for the Indian millennial who has been accustomed to 4G broadband speeds. However, characteristics such as "typing is preferred to writing," when extended to Indian youth have become, 'voice is preferred over typing' (Bellman, 2017). This behavior will speak to how they are going to use the language while browsing and texting. Traditional forms of English instruction and assessments are therefore in need of a reformation to meet the millennial learners' needs.

Blom and Saeki (2011) Pandey and Pandey (2014) have conducted studies measuring the effectiveness of English Language courses offered in colleges in India and they found them to be failing the students. Engineering students who make up a considerable portion of graduates entering the Indian job market each year, have commented "Written exams must be abolished for English courses. More innovation is required in the methodology for creating enthusiasm" (Clement & Murugavel, 2015). The major reason for the non-employability of the majority of Indian graduates is the absence of effective communication skills (Aspiring Minds [2014]. National employability report – Engineers. Annual report – 2014). Moreover, the inability to speak in a language that was the primary medium of instruction for the majority of their education leaves them doubting their abilities and with low self-confidence.

English Needs of Professional Education Students

Based on the survey of professional education students like students of Engineering, Law, and Teacher Education programs, their needs to learn the English language were categorized into five divisions;

1. Aural Oral skills-related Needs
2. Reading and Writing skills-related to Needs

3. Academic-related Needs
4. Employment-related Needs, and
5. Social-related Needs.

Aural Oral Skills-Related Needs

Students give importance to oral skills like listening and speaking more than reading and writing skills. Not only in the learning environment, but also for social purposes, placement values, and establishing one's own identity in a scholarly environment, spoken language is highly preferred to written. Speaking English fluently is considered the topmost priority among students. Next to that responding in English during the classroom interaction got the second most priority. English is needed to listen to news on TV or Radio and students need to speak English with the proper voice modulation and intonation. Next to that, students need to understand English while watching English programs in movies and to communicate in English and speak English without any grammar mistakes. The pandemic situation of COVID-19 encourages self-learning and self-determined learning (Heutagogy) among students; it paves the way for lifelong learning and exploring technological resources like videos and blogs. Hence, the need to understand English while listening to academic videos and podcasts is also on the priority list. Acquiring native-like pronunciation and the need to understand the teacher's lectures in the classroom interaction is considered the twelfth priority. Very few students expressed that knowing native speakers' use of English with their accent, grammar, and vocabulary is their priority.

Reading and Writing Skills-Related Needs

To quote Francis Bacon, 'Reading maketh a perfect man'. Among the four skills, reading makes one a self-dependent and a life-long learner without depending on others. Once the reading competency is properly cultivated among the students, they will become self-reliant throughout their lifetime. Students consider reading and replying to emails in English as the top most needed skill in this second dimension of the survey, as most of the official and formal communication takes place online. Next to that 'writing answers in English without grammar mistakes in exams' is considered as important. When it comes to writing, accuracy is deemed as more important. Hence, the students prefer their writing as a scholarly one. Following this, the ability to read and understand the books related to their academic subjects is considered as the next important need. Next to that, four items are considered equally by the students as the third priority in their list. They are 'knowledge of English to read newspapers, reading aloud the text with proper pronunciation and following the punctuation marks, the knowledge of spelling and the need to read the research articles for the studies related to their academic subjects'.

To sustain this interest in reading among students, teachers create a suitable learning exposure and provide appropriate guidance in fixing the right choice of reading material. For further facilitation, study skills and reference skills should also be nurtured among students. Reading nowadays is not only confined to reading books and other materials from the library but also exploring the Open Educational Resources (OERs). Hence, the teachers have to orient them on how to extract the needed information from online sources and how to curate the content in depth.

Academic-Related Needs

The need for an evaluation scheme to test the student's compositional skills is the topmost priority of the students. As the students are fed up with memory-based learning, they opt for a constructivist approach for both learning and testing. Instead of depending on the teacher-given material or bazaar guides, the students want to write on their own. To satisfy this need, the summative evaluation alone is not sufficient. Along with the content of the subject course, all faculty members must correct the language errors of the students. Formative evaluation was one of the solutions. Instead of assessment of learning, assessment should be done 'for' learning, and it would be still better if the assessment itself functions 'as' learning process. Students' compositional skill has to be encouraged from the beginning scientifically. The third statement in their priority list is the need for learning resources in the form of books, audio/video materials, and language lab facilities. In this digital age, apart from printed materials, students are after electronic resources like videos, audio, modules, and language laboratories. It is high time for the academic administrators to facilitate the learners with the needed learning infrastructure. The next concept on the priority list is the assessment of listening and speaking skills in the university exam paper. This wish throws light on the learners' focus on the development of oral skills on par with writing. The modification of syllabi by giving equal importance to all four communication skills is also highlighted by the learners.

Employment-Related Needs

The need to write resumes and applications for jobs in English is considered the highest need in this dimension. The immediate need of any graduate is employment in the proper concern/institution. Next to that comes the need to have a good command of English to face job interviews. Appearing for the interviews, one needs to be equipped not only with verbal communication but also paralinguistic elements of communication. The third priority is the need for English to fetch good and attractive careers to students. Students also express their need to access more learning resources for further professional development. The need for English to crack competitive exams is considered as the least priority among the other priorities.

Social-Related Needs

The need for English to use online social network sites is strongly felt by the students as it is the order of the day for everyone to be a member of any one of the social networking sites. To follow the celebrities and icons of various fields like sports, youngsters need to be familiar with English. Students consider the need for the English language to attain a high social status in the country. Of course, no one can deny the craze of English in Indian society. Still, English is considered the language of high status and rank.

Implications

i. **Focus on Oral Skills:**

   The teachers of English are expected to focus more on improving the students' listening and speaking skills as they later consider these oral communication skills as their primary need. Fluency and following other phonetics aspects like speaking with proper stress and voice modulation are felt more important by the students which implies that the teachers should concentrate more on teaching them. The best way to teach is to model it. The teachers of English need not rely on delivering the lectures and dealing with the syllabi alone. Instead, they may play some academic-related audio/videos by native speakers and other eminent personalities and thereby the teachers can create an English ambiance. It is not only the transformation of information and comprehension but training in listening and speaking for which continuous and planned instructional design must be prepared by the teachers. They may use real-life-oriented resources like radio and television (English channel) and ICT tools like podcasts, sending audio files through LMS and YouTube videos for language improvement of the students. Teachers should make it compulsory to have an English atmosphere at least in the English classrooms.

ii. **Extended Reading and Functional Writing Skills:**

   The primary need of the students related to reading and writing skills is to write without spelling mistakes. When it comes to higher education, teachers overlook the importance of teaching spelling. Moreover, due to the advancement of artificial intelligence, spelling errors can be found out and corrected on the screen. But it is expected to introduce the new words along with their spelling. As far as reading is considered, students prefer to read extensively like reading-related reference materials, research articles, and newspapers.

Hence, the teachers of English are expected to extend the selection of reading materials adding the recent research and other materials along with the prescribed texts. The students' responses throw light on the need to use online written communication which implies that the teachers have to teach the functional writing skills of the students for common and professional communication. The etiquette of online writing (like e-mails) needs to be added to the instruction.

iii. **Acknowledging the own composition**

The digital age students are fed up with repeating what they read, instead, they want to compose on their own. The scope to express their understanding in their language should be created by the teachers. Assessment strategies should consider the oral examination more than the paper-pencil test, which will also be more reliable in testing the authentic comprehension skills of the students. The lowest score (3.97) for the statement 'The testing and assessment schemes of the present day meet the needs of students.' implies that the teachers have to utilize continuous and comprehensive evaluation strategies considering the development of all skills equally.

iv. **Consideration of Employability skills:**

The teacher's duty does not end with the curriculum transaction but extends to boost up the employability skills of the students too. Language teachers should also train the students to develop their soft skills. They should guide them on the preparation of a good resume, interview skills, and how to face competitive exams. They have to sail with the tide and keep updated with the recent trends in business communication. Apart from subject-related explanations, the teachers have to give them exposure to functional situations like introducing themselves, negotiating with others without hurting their sentiments, agreeing with others, polite disagreement, summing up a discussion, conversing with a sense of humor in needed places, cultural differences in communication, non-verbal communication, etc. which will elevate the students' disposition both in the interview and during the employment period.

v. **Comprehensive Approach:**

Teachers have to shift the focus from writing skills alone to all the other skills. Equal concentration on listening, speaking, and reading is needed for a good command of the English language.

Conclusion

The research findings pointed out the students' expectations as follows; (i) Increased concentration on fluent and effective communication skills rather than reading and writing, (ii) Encouraging the students' written compositional skills (iii) Developing their employability skills by honing the soft skills considerably, (iv) Effective integration of online and ICT tools while teaching, (v) Building up the confidence of speaking in English by creating and nurturing English ambiance.

References

Bellman, E. (2017, August 7th). The end of typing: The next billion mobile users will rely on video and voice. *The Wall Street Journal.* Retrieved from https://www.wsj.com/articles/the-end-of-typing-the-internets-next-billion-users-will-use-video-and-voice-1502116070

Blom, A., & Saeki, H. (2011). Employability and skill set of newly graduated engineers in India. *Policy Research Working Paper* (p. 5640). World Bank, Washington, DC, 1–33.

Clement, A., & Murugavel, T. (2015). Professional development of English professors in Indian engineering colleges: The need of the hour. *English Language Teaching*, 8(5), 132–142. https://files.eric.ed.gov/fulltext/EJ1075200.pdfPandey, M., & Pandey, P. (2014). Better English for better employment opportunities. *International Journal of Multidisciplinary Approach and Studies*, 1(4), 93.

Chapter 7

A Systematic Literature Review on Student Engagement in e-learning

Challenges and Prospects in Higher Education

Raj Kumar Pal, Nil Ratan Roy, and Antara Dey

Introduction

E-learning is now a compulsion across the globe at any level of education. E-learning has many names like online learning, distance learning, and virtual learning. It has many types and forms, for example, massive open online courses, virtual learning environments, and learning management systems. E-learning provides a new way to teach students, challenges established learning methods, and draws students from all over the world. With the advancement of digital technologies, e-learning has become more popular and common to people. The reason is that it has a unique feature of overcoming time and space constraints on learning. It also enables students to pursue online educational courses while performing their societal responsibilities. Because of the features of online learners, dropout rates in online modules have been higher than in traditional education. Recognizing low levels of engagement in online learning, different educational institutions have taken the initiative to identify problems and research possible solutions for engaging students in online courses.

Digital technologies in the form of the internet deliver flexibility and convenience in the collection and exchange of information. It offers students a wealth

DOI: 10.1201/9781003504894-7

of online learning materials, a plethora of online learning tools, and a wide variety of online learning environments, knowledge representation, and styles of presentation. As a result, online learning has progressively become an essential mode of instruction. Without rigorous criteria for certain demographic features like the age of the learners, their educational background, and social status, they are free to choose appropriate learning content according to their needs, and interest and can improve their learning progress flexibly throughout online learning. This is one of the prominent causes of the popularity of e-learning among students. During the time of social isolation caused by the COVID-19 epidemic, the modes of education have moved from face-to-face education to completely virtual modes of education, which are heavily reliant on contemporary technology. Such transitions have resulted in a gap between social interaction among learners and online education and work. Assessing student motivation and interaction is an important component of managing the learning environment.

E-learning is the use of an electronic medium for asynchronous or synchronous communication for knowledge development. Organizations are increasingly embracing e-learning due to its uniform global training, decreased delivery cycle time, enhanced learner convenience, less information overload, improved tracking, and cheaper costs (Welsh et al., 2003). Asynchronous learning solves the problem of multiple time zones and distance for learners by allowing them to learn whenever and wherever they choose (Ally, 2004). E-learning has been applied successfully in academic institutions and documented improvements in teaching and learning quality, as well as improved revenue, learning outcomes, and satisfaction (Chang, 2016). E-growing learning's popularity might also be due to the flexibility it provides at a lower cost. Furthermore, the fourth aim of the United Nations Sustainable Development Goals stresses quality technical, vocational, and tertiary education by 2030, and advocates that e-learning may emerge as a tool to take care of several major issues in the context of higher education such as access, quality, and cost (Yeld, 2016).

Conceptual Background

Concept of e-learning

E-learning is the application of sophisticated information and communication technology in the learning process, with advanced technology referring to electronic media (Thomas & Graham, 2019). It is a new educational pattern forced by new scientific and technological developments occurring in the world today as a result of the incapacity of conventional teaching tactics and methods to keep pace with these changes, given the shifting roles of both teacher and student (Gorbi, 2013; Harandi, 2015). E-learning is currently regarded as one of the most significant educational settings in the digital era, according to Zare et al. (2016). The globe has experienced upheaval and quick developments in information technology

(ICT) and the internet in many sectors, notably in the education sector, throughout the last century. This has caused educational institutions and curriculum specialists to implement dramatic modifications in their policy and academic attainment to adapt to and conform to this shift and development (Choshin & Ghaffari, 2017; Lim & Morris, 2009; Vella-Brodrick & Klein, 2010).

E-learning is the use of electronic technology for educational curricula beyond the confines of a traditional classroom, and it has grown in popularity, particularly among the younger population, during the last decade (Singh & Mishra, 2019). E-learning is the international application of networked information and communication technologies in teaching and learning. Online learning, virtual learning, distributed learning networks, and web-based learning are examples of these.

Concept of Student Engagement

One of the qualitative measures in the learning process is engagement (Jang, Park, & Yang, 2014). Student involvement has progressed from a single dimension to a multidimensional process, with early studies focusing on the behavioral dimension, followed by the behavioral and emotional dimensions, and ultimately joining the cognitive component. The concept is made up of three major components: behavioral engagement (staying on task), emotional engagement (boredom), and cognitive engagement (focused attention). Participation is derived from the Latin "*particpare*"; nevertheless, in the literature and research, the terms engagement, involvement, and participation are frequently employed, with engagement being the most commonly used. Researchers proposed student involvement as a way to integrate the notion of participation into the area of education. Student involvement, learning involvement, and learning participation are other terms for student engagement.

Theoretical Perspectives

The theoretical perspectives of student engagement help to conceptualize the nature of the student engagement construct. Student engagement-related problems and concerns may be mitigated by devising better control over the e-learning environments. The theories substantially help in providing directions to plan remedies for the problems regarding e-learning. Two theoretical viewpoints were chosen to help the colleges better handle low online engagement and attrition. For starters, critical literacy may assist in verifying the assumptions that underpin traditional organizational commitments to rising students' diverse needs and changing technological evolution. It entails a paradigmatic shift from old beliefs in which students are viewed as having a "deficient" if they are unable to properly navigate academic surroundings. Critical Discourse Theory recasts students' experiences as 'unfamiliar,' acknowledging students' creative and innovative ideas and attitudes that may not meet the institutional goals. Second, Communication theory also supports the critical discourse theory that learning and teaching are fundamentally communication processes that ensure effective communication of students with teachers and peers, which eventually assist the student engagement practices in the virtual learning environment.

Self-determination theory is a macro-level theory of individuals' motivation that explains the dynamics of need, motivation, and welfare of the learners in their social environment (Deci & Ryan, 1985). The theory also highlights that all persons have three fundamental and psychological demands that motivate them to act or not act: autonomy of behavior, competence in any activity, and relatedness to other social members. Individuals experience better psychological well-being when these three psychological requirements are satisfied, and when these needs are not met, they feel excessively fragmented, lonely, and reactive. According to Hsu et al. (2019), students are intrinsically motivated to actively participate in learning activities when pedagogical design effectively satisfies these psychological demands. Reeve (2013) showed that learning environments that address the three fundamental psychological demands have a higher likelihood of engaging pupils in learning. As a result, this theory helps explain how needs-based assistance affects student motivation, engagement, and learning.

Objectives of the Study

The main objectives of the present study were to identify the problems encountered by students when engaging in an e-learning environment, to highlight the challenges faced by instructors when engaging students in effective online learning, and to suggest how e-learning can be made more effective in engaging students.

Methods and Procedures

We performed a systematic review of the available literature on student engagement in e-learning. For this chapter, we followed PRISMA guidelines (Moher et al., 2009) to select the articles for systematic review. The detailed procedures of literature review have been discussed here (Figure 7.1):

Stage I – Planning the Review

Search Strategy

Firstly, we planned to collect scholarly articles on student engagement in the e-learning environment. We searched for research articles on the following databases: (1) Scopus, (2) Google Scholar, (3) Elsevier/Science Direct, and (4) IEEE.

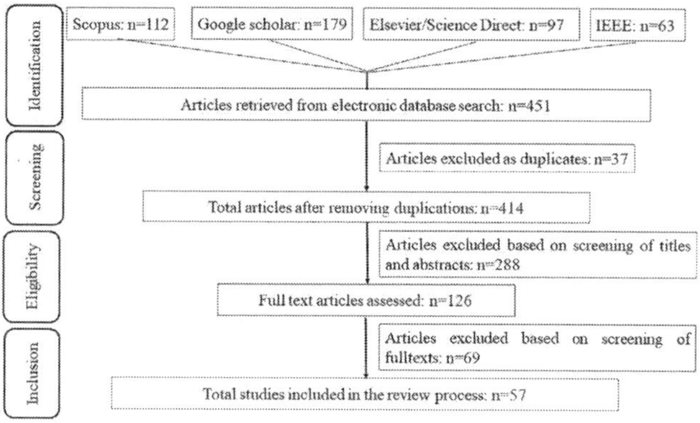

Figure 7.1 Details of the procedures followed while conducting the literature review.

Search Terms

We started with the following combinations of the search strings using appropriate Boolean operators as follows:

(Student engagement OR student involvement OR student participation)AND (E-learning OR online learning OR virtual learning OR open learning)

The syntax of the keywords was adjusted accordingly for each database.

Identification of Relevant Literature

After considering the results from different databases, the articles that were written in any language other than English were removed. Further, the duplicate articles were also identified and flagged based on title, publication year, and digital object identifier (DOI). We excluded all the articles which were published before 2017 from this list. Eventually, a list of 451 articles was obtained for conducting the review process. After reviewing the literature, a large number of studies were found. The researchers delimited this study by reviewing articles published in the last five years. The timeline was set considering the scope of the article into consideration. Additionally, for the same reason, the platforms namely, Scopus, Google Scholar, Elsevier/Science Direct, and IEEE Access were also selected purposively.

Stage II – Conducting the Review

All the articles obtained at stage I were screened and examined for their fit with the main theme of the present study. The screening was done primarily by considering

Table 7.1 Details of the Inclusion and Exclusion Criteria for the Published Articles in the Process of Literature Review

Inclusion Criteria	*Exclusion Criteria*
Research papers that were published between 2017 and March 2021.	Articles published before 2017 and after March 2021.
Research papers published on well-known databases like Scopus, Google Scholar, Elsevier/Science direct, and IEEE Access.	Articles published on non-academic databases.
Research articles that were written in English	Articles compiled in any language other than English.
Articles with open access and with available full text. Articles that have discussed problems and challenges of the e-learning environment.	Books and grey literature like doctoral and master's dissertations, project reports, conference articles, working reports, etc.

the title, abstract, and keywords of each article and necessarily by reading the full-length paper.

Inclusion and Exclusion Criteria

The inclusion and exclusion criteria were set to restrict the search results for relevant articles that fall under the scope of the current study (see Table 7.1). After applying the pre-determined inclusion and exclusion criteria, we obtained a sample of 57 articles that were published in 27 scholarly journals.

Stage III – Reporting and Dissemination

A comprehensive review of the articles in the final sample obtained at stage II was conducted. The results were organized and presented to fulfill the objectives of the present study.

Data Extraction

We reviewed and summarized each remaining article at this point, gathering the following information on study topics, data utilized, methods, variables, and conclusions. We compiled this data into a preliminary report and all the authors together discussed the findings and implications. We used the outcomes of the discussion to modify the report.

Results and Findings

Problems of the Students in the e-learning Environment

In terms of technical skills, online learning and changing to a new manner of study can present several challenges for college students, and the anxiety associated with using online learning platforms can harm learners' performance and confidence. The shift to online delivery reduces information transmission and the chance for students to put what they've learned into practice, while the instructor's role shifts from teaching to aiding with the learning process. When this happens, the online learning environment, especially if there is no sense of belonging or peer engagement, becomes an inadequate medium for information transmission, and it must be reinforced by students doing peer evaluation.

Although the transition from a face-to-face learning environment to online learning may be beneficial for some learners who previously viewed it as a flexible medium for delivering educational programs' anytime, anywhere,' it can pose difficulties for learners in different countries when an online course spans multiple time zones. Students may fall asleep in class or change their weekend and holiday plans as a result, making it more difficult for them to be fully involved in their online education. Furthermore, learning online has certain health risks, including eye strain from reading e-books or working on digital devices, as well as general exhaustion.

Due to the likely absence of social indications such as facial expressions and eye contact in online lectures, students' social participation may deteriorate. Meanwhile, human connection develops when students use their video conference cameras to view each other in an online learning situation.

Another issue with e-learning is the lack of social presence, which causes students to feel isolated as a consequence of a lack of peer communication, which can lead to students dropping out of the course and negatively impacting learning. Students can cope with the sense of being left behind by engaging in social contact. In online classes, student-to-student interaction is viewed as critical to enhancing student engagement and retention. One way to solve the problem is to strengthen the bonds between members of a group and increase their sense of belonging. The word "sense of belonging" relates to students' feelings of identification and affiliation with others, which contributes to their motivation to study. A sense of belonging is a mediating variable that explains how a peaceful classroom atmosphere is achieved. Student participation in online learning is influenced by a harmonious classroom environment (good student-teacher and student-student relationships).

Although quantitative and mixed methods are utilized in research, qualitative methods like questionnaires and self-reports are more popular. Student engagement is frequently quantified in engagement research for quantitative studies by assessing students' behavior and actions in the virtual learning environment. However, judging student participation only based on their conduct is imprecise and misleading.

Most researchers looked at student engagement as a whole, rather than looking at individual parts, of the study item.

Challenges towards the Instructors in Engaging Students in e-learning

Customization is one of the major challenges for e-learning systems and methods (Shailaja & Sridaran, 2014). As everyone learns differently, it is quite difficult to cater individual needs of each learner. However, it is quite important to design the instructions as per the interest of the learners, adjust teaching strategies, provide importance to learner autonomy and preferences, and personalize the contents. An engaging and motivating learning environment must be properly planned by the instructors to dissuade students from becoming anxious and unrest when they feel that they are learning slower than other students especially when there is a lack of teacher-student or student-student interactions.

Because of the challenges in maintaining monitoring and control qualities, distant learning may harm students' self-confidence and self-efficacy. The virtual learning environment is a modern approach that uses digital technologies to record students' learning behavior and their reactions while involved in various learning activities. It also evaluates students' emotional and behavioral aspects of learning.

Online students complete their courses at a lower rate than students involved in offline learning modes (Eastin & LaRose, 2000). The study also emphasizes how to effectively manage learners' learning needs and expectations, obtain quality educational outcomes eliciting a higher level of student involvement in the courses, and improve students' self-efficacy, investment of time, and self-confidence in accomplishing desired learning objectives (Carbone et al., 2019). According to this study, there is uncertainty over whether students understand how to utilize technology-based tools strategically to enhance their learning experiences, as Burton et al. (2015) also argue.

Discussions

Student engagement is a multifaceted notion that encompasses student conduct, cognition, and emotion, as well as explicit behavior. For the measurement of such a multi-dimensional construct, any standalone method is insufficient. For its effective measurement, a mixed approach i.e. the quantitative methods as well as the qualitative methods may be applied together to comprehensively capture students' engagement learning processes. This will provide a detailed measurement of each engagement dimension. The output of such measurement will be more effective in formulating practical measures to enhance the levels of engagement in the required dimensions. This becomes a more effective process to ensure the quality learning outcomes of the students in the online environment.

Measurement of Student Engagement

Three prominent measures of student engagement include self-reports, observation schedules/checklists, and automated instrumental measures. Both self-report instruments and observation schedules/checklists are in vogue and are unsuitable for real-time learning systems. Currently, affective computing may be used to automatically assess engagement. Smart systems for detecting students' emotions have been developed in the subject of affective computing. Some studies emphasize the significance of monitoring learners' feelings and the methods for detecting those as a reflection of student involvement. The readings of physiological and neurological sensors, such as electroencephalography, heart rate, and facial expressions are used to automate measurements.

Based on our observations, we will have additional possibilities to improve the accuracy of engagement measures whenever multiple modes of measurement techniques are applied. Education in an e-learning environment and student engagement has piqued the interest of experts to improve studies in the fertile area of engagement research. The plausible causes behind this may be as follows: (1) The measurement of the multi-dimensional construct, i.e. student engagement is critical in developing online learning systems, thus an accurate automated method is required to improve the learning process. (2) Current techniques for detecting involvement that depend just on emotions perform poorly. They are unsuitable for real-time systems because they employ costly technologies and violate student privacy.

Online Learning Analytics

The learning management systems record students' behaviors during online learning and provide useful feedback to the course instructors regarding teaching, learning, assessment of learning, and other educational decision-making. The learning analytics captures students' access to educational resources and usage. The course instructors have access to detailed information regarding the performance and participation of the learners in an online course. This may be useful in improving teaching strategies and the instructors may get ideas for adjusting their teaching by formulating and re-designing instructions and communication with students.

There are creative practices throughout the world that use learning analytics to record learners' engagement in the online learning environment, offer a real-time scenario of student involvement, and track students' behavior that reflects lower engagement. Tracking statistics provides detailed information about students' activity even at individual levels, for example, when and how frequently individual students accessed the supplied learning resources, details of submission of homework and assignments, and all the records of previous assessments. This helps instructors to provide feedback and judgments regarding targeted support or early intervention. Even though there has been a significant study on institution-wide analytic-tracking treatments and their assessments, fewer studies have been carried out on course-specific learning analytics and how they may be used to improve student engagement.

Student Engagement Strategies

One of the issues with online learning appears to be intimidation, particularly among students who are new to such a learning environment, and a lack of trust among peers, which results in lower participation in discussions and debates. Intimidation can also occur when students believe their peers know more about the topic at hand or are dominating the online class discussion.

Increased online classroom engagement relies heavily on group cohesion and mutual trust. Large group sizes and numerous time zones can disrupt the learning process and cause group incoherence in synchronous classrooms, whereas small class sizes improve "sociability, social space, and group cohesiveness" (Akcaoglu & Lee, 2016, p. 13).

To better manage discussion threads, Woods and Bliss (2016) advocated forming smaller working groups, which would include reflective tasks that would promote peer participation and provide timely constructive feedback to facilitate online student debate. Peer monitoring is beneficial to students who participate in online group discussions.

Remedies to Low Student Engagement in e-learning:

The following may be suggested to cope with the existing problems and concerns regarding students' engagement in the e-learning platforms:

Creating and Sustaining a Pleasant Learning Environment

The motivation of students is strongly related to their learning experiences. As a result, motivation may be the key to gaining students' attention. External variables such as deadlines and prizes may sometimes be harmful to cognitive capacity, creativity, and problem-solving. The authors also claimed that encouraging a healthy competitive atmosphere boosted the motivation of students. Furthermore, encouraging critical thinking in a secure learning environment may result in favorable outcomes. Learners can be prompted to think critically through debating, assessing, and making conclusions about subjects.

Giving Consistent Feedback Promptly

All of the studies analyzed advised that the best engagement tactics include delivering timely feedback to online students and ensuring that they are engaged throughout the course. Also, rather than giving them the solution immediately away, kids occasionally want assistance in obtaining the answers. Constructive input from professors and peers is always valued by students. Extrinsic motivation may also be created through positive feedback, compliments in the form of acknowledgment for good work, and critical remarks.

Using the Proper Technology to Provide the Correct Content

Using a learning management system (LMS) creatively can result in learner engagement. Even in huge online courses, effective use of existing technology may increase teaching quality. To encourage contact between the teacher and fellow students, methods like as Skype and Blackboard collaboration can be employed. Additionally, e-mail and synchronized and asynchronous chat sessions can help learners stay in the loop and engaged.

Providing a Good Support System

Previous study has demonstrated the value of providing students with the appropriate resources. When planning and delivering online course content, teachers must be cautious about using the appropriate technologies. Library, databases, and Turnitin (to deal with plagiarism) are examples of organized and easy-to-access tools that may help learners feel competent in the classroom. However, due to a lack of computer skills and technological issues, learners in online courses may not be able to fully utilize the offered materials. They suggested that open contact with students and tracking their actions via a learning management system (LMS) can assist in addressing the problem.

Conclusions

Firstly, the instructors play a pivotal role in the online discourse of learning. The instructors must be aware of the impact of student engagement on their academic output which is reflected through their performances and academic achievements. Further, the different engagement levels are to be realized and identified so that the students with lower engagement may be marked. Hence, it is quite important to know about the different engagement levels of the students and accordingly identify students in each level. It can be noted that the course contents and instructional techniques are difficult to separate from the instructor. Secondly, social relatedness must be encouraged in online learning through teacher-student and student-student interactions. The intrinsic motivation of students must be coupled with the sustained encouragement of the instructors. This will strengthen the emotional attachment of the learners and also affect the cognitive engagement dimension. Thirdly, teachers must improve their digital communication skills to enrich their online teaching strategies. They should continuously think about how to formulate an interesting and innovative teaching-learning environment. This will boost the levels of emotional engagement among the students and reciprocally the teachers will also be motivated to adjust their teaching strategies. The instructional designs should be focused on relatedness between students and teachers.

Teachers and educational institutions should take the initiative to promote students' participation by supporting relatedness in the online courses. Students are to be provided ample space so that they can interact with the instructors and their peers. This will develop a sense of belongingness among the students. The course instructors may arrange some mental activities so that the mental fatigue of the students may be reduced. This will help to sustain their concentration level and help them to avoid boredom. Further, the instructors may design their instruction by creating a social environment in learning by introducing online peer support groups, adopting innovative assessment approaches, and incorporating interdisciplinary teaching strategies and interesting learning activities. Finally, educational institutions should consider screen fatigue and provide adequate training and break periods to allow students and teachers to maintain their mental health throughout the semester while taking online courses.

Bibliography

Akcaoglu, M., & Lee, E. (2016). Increasing social presence in online learning through small group discussions. *The International Review of Research in Open and Distributed Learning, 17*, 121–135.

Ally, M. (2004). Foundations of educational theory for online learning. *Theory and Practice of Online Learning, 2*(1), 15–44.

Almarashdeh, I. (2016). Sharing instructors' experience of learning management system: A technology perspective of user satisfaction in a distance learning course. *Computers in Human Behavior, 63*, 249–255.

Anderson, T., & Garrison, D.R. (1998). Learning in a networked world: New roles and responsibilities. In *Distance Learners in Higher Education: Institutional Responses for Quality Outcomes*. Atwood(pp. 97–112). http://hdl.handle.net/2149/801.

Archer, M. S. (2003). *Structure, Agency, and the Internal Conversation*. Cambridge University Press.

Arnold, K. (2010). Signals: Applying academic analytics. *Educause Quarterly, 33*(1), 10–21.

Avella, T. J., Kebritchi, M., Nunn, G. S., & Kanai, T. (2016). Learning analytics methods, benefits, and challenges in higher education: A systematic literature review. *Online Learning Journal, 20*(2), 13–29.

Bandura, A. (1986). *Social Foundations of Thought and Action*, 1st ed. Prentice-Hall.

Bickle, M. C., Rucker, R. D., & Burnsed, K. A. (2019). Online learning: Examination of attributes that promote student satisfaction. *Online Journal of Distance Learning Administration, XXII*(1), 1–7.

Broadbent, J. (2017). Comparing online and blended learners' self-regulated learning strategies and academic performance. *The Internet and Higher Education, 33*, 24–32.

Burton, L. J., Summers, J., Lawrence, J., Noble, K., & Gibbings, P. (2015). Digital literacy in higher education: The rhetoric and the reality. In M. K. Harmes, H. Huijser, & P. A. Danaher (Eds.), *Myths in Education, Learning, and Teaching: Policies, Practices, and Principles* (pp. 151–172). Palgrave Macmillan.

Carbone, A., Drew, S., Ross, B., Ye, J., Phelan, L., Lindsay, K., & Cottman, C. (2019). A collegial quality development process for identifying and addressing barriers to improving teaching. *Higher Education Research & Development, 387*(7), 1–15. https://doi.org/10.1080/07294360.2019.1645644

Castaño-Muñoz, J., Colucci, E., & Smidt, H. (2018). Free digital learning for inclusion of migrants and refugees in Europe: A qualitative analysis of three types of learning purposes. *The International Review of Research in Open and Distance Learning, 19*(2), 1–21. https://doi.org/10.19173/irrodl.v19i2.3382

Chang, V. (2016). Review and discussion: E-learning for academia and industry. *International Journal of Information Management, 36*(3), 476–485.

Chen, B., Bastedo, K., & Howard, W. (2018). Exploring design elements for online STEM courses: Active learning, engagement & assessment design. *Online Learning, 22*, 59–75.

Choshin, M., & Ghaffari, A. (2017). An investigation of the impact of effective factors on the success of e-commerce in small and medium-sized companies. *Computers in Human Behavior, 66*, 67–74.

Czerkawski, B. C., & Lyman, E. W. (2016). An instructional design framework for fostering student engagement in online learning environments. *Tech Trends, 60*, 532–539.

D'Errico, F., Paciello, M., & Cerniglia, L. (2016). When emotions enhance students' engagement in e-Learning processes. *Journal of e-Learning and Knowledge Society, 12*. https://www.learntechlib.org/p/173676/

Deci, E. L., & Ryan, R. M. (1985). *Intrinsic Motivation and Self-determination in Human Behavior*. Plenum Press.

Delone, W. H., & McLean, E. R. (2003). The DeLone and McLean model of information systems success: A ten-year update. *Journal of Management Information Systems, 19*(4), 9–30.

Eastin, M. S., & LaRose, R. (2000). Internet self-efficacy and the psychology of the digital divide. *Journal of Computer-Mediated Communication, 6*(1), 589–611. https://doi.org/10.1111/j.1083-6101.2000.tb00110.

Fredricks, J. A., Blumenfeld, P. C., & Paris, A. H. (2004). School engagement: Potential of the concept, state of the evidence. *Review of Educational Research, 74*(1), 59–109.

Garratt-Reed, D., Roberts, L. D., & Heritage, B. (2016). Grades, student satisfaction and retention in online and face-to-face introductory psychology units: A test of equivalency theory. *Frontiers in Psychology, 7*(673), 1–10. https://doi.org/10.3389/fpsyg.2016.00673

Garrison, D. R. (2011). *E-learning in the 21st Century: A Framework for Research and Practice*. Taylor and Francis.

Gorbi, A. (2013). *The Investigation of Relationship between Information Communication and Technology (ICT) and Critical and Creative Thinking Skills* (Master's Thesis, Islamic Azad University).

Harandi, S. (2015). Effects of e-learning on students' motivation. 3rd International Conference on Leadership, Technology and Innovation Management. *Procedia-Social and Behavioral Sciences, 181*, 423–430. https://www.sciencedirect.com/science/article/pii/S1877042815031985.

Hsu, H. C. K., Wang, C. V., & Levesque-Bristol, C. (2019). Reexamining the impact of self-determination theory on learning outcomes in the online learning environment. *Education and Information Technologies, 24*(3), 2159–2174. https://doi.org/10.1007/s10639-019-09863-w

Hu, P. J. H., & Hui, W. (2012). Examining the role of learning engagement in technology-mediated learning and its effects on learning effectiveness and satisfaction. *Decision Support Systems, 53*(4), 782–792.

Hussein, R., & Hassan, S. (2017). Customer engagement on social media: how to enhance continuation of use. *Online Information Review, 41*(7), 1006–1028.

Jang, M., Park, C., & Yang, H. S. (2014). Building an automated engagement recognizer based on video analysis. In *9th ACM/IEEE International Conference on Human-Robot Interaction (HRI)*, pp.182–183.

Jordan, K. (2015). *MOOC Completion Rates: The Data.* http://www.katyjordan.com/MOOCproject.html

Jung, Y., & Lee, J. (2018). Learning engagement and persistence in massive open online courses (MOOCS). *Computers & Education, 122*, 9–22.

Kankanhalli, A., Pee, L. G., Tan, G. W., & Chhatwal, S. (2011). Interaction of individual and social antecedents of learning effectiveness: A study in the IT research context. *IEEE Transactions on Engineering Management, 59*(1), 115–128.

Kim, C. M., Park, S. W., Cozart, J., & Lee, H. (2015). From motivation to engagement: The role of effort regulation of virtual high school students in mathematics courses. *Journal of Educational Technology & Society, 18*(4), 261–272.

Kori, K., Pedaste, M., Altin, H., Tonisson, E., & Palts, T. (2016). Factors that influence students motivation to start and to continue studying information technology in Estonia. *IEEE Transactions on Education, 59*(4), 255–262.

Kushwaha, R. C., Singhal, A., & Chaurasia, P. K. (2015). Study of students' performance in learning management system. *International Journal of Contemporary Research in Computer Science and Technology (IJCRCST), 1*(6), 213–217.

Kuzilek, J., Hlosta, M., Herrmannova, D., Zdrahal, Z., & Wolff, A. (2015). OU analyses: Analyzing at-risk students at the Open University. In *Proceedings of First International Workshop on Visual Aspects of Learning Analytics and Knowledge Conference*, pp. 1–16, Poughkeepsie, NY, USA.

Lee, E., Pate, J. A., & Cozart, D. (2015).Autonomy support for online students. *Tech Trends, 59*(4), 54–61.

Liai, L., & Qinglin, Z. (2003). The types and characteristics of online learning. *Education Science, 1*, 49–52.

Lim, D. H., & Morris, M. L. (2009). Learner and instructional factors influencing learning outcomes within blended learning. *Journal of Educational Technology & Society, 12*(4), 282–293.

Martin, F., & Bolliger, D. U. (2018). Engagement matters: Student perceptions on the importance of engagement strategies in the online learning environment. *Online Learn, 22*, 205–222.

Meyer, K. A. (2014). Student engagement in online learning: What works and why. *ASHE Higher Education Report, 40*(6), 1–14. https://doi.org/10.1002/aehe.20018.

Min, H., Hao, L., Wenping, D., & Hua, G. (2016). Student engagement: One of the necessary conditions for online learning. *International Conference on Educational Innovation through Technology, 1*, 122–126.

Mohammadi, H. (2015). Investigating users' perspectives on e-learning: An integration of TAM and IS success model. *Computers in Human Behavior, 45*, 359–374.

Moher, D., Liberati, A., Tetzlaff, J., & Altman, D. G. (2009). Preferred reporting items for systematic reviews and meta-analyses: The PRISMA statement. *PLoS Med, 6*(7). Retrieved from https://doi.org/10.1371/journal.pmed.10000 97.

Murray, M. C., Pérez, J., Geist, D., & Hedrick, A. (2012). Student interaction with online course content: Build it and they might come. *Journal of Information Technology Education: Research, 11*, 125–140.

Nelson, K., & Creagh, T. (2012). *A Good Practice Guide: Safeguarding Student Learning Engagement.* The Queensland University of Technology.

Palloff, R. M., & Pratt, K. (2001). *Lessons from Cyberspace Classroom.* Jossey-Bass.

Paulsen, J., & McCormick, A. C. (2020). Reassessing disparities in online learner student engagement in higher education. *Educational Researcher, 49*(1), 20–29. https://doi.org/10.3102/0013189X19898690

Perrotta, C., & Williamson, B. (2018). The social life of learning analytics: Cluster analysis and the 'performance' of algorithmic education. *Learning, Media and Technology, 43*(1), 3–16. https://doi.org/10.1080/17439884.2016.1182927

Poon, A., Giroux, S., Eloundou-Enyegue, P., Guimbretière, F., & Dell, N. (2020). Baccalauréat practice tests in Cameroon: The impact of sms-based exam preparation. In *Proceedings of the 2020 International Conference on Information and Communication Technologies and Development*, Guayaquil, Ecuador, 17 June 2020; pp. 1–12.

Radovan, M. (2019). Should I stay, or should I go? Revisiting student retention models in distance education. *Turkish Online Journal of Distance Education, 20*(3), 29–40.

Rashid, T., & Asghar, H. M. (2016). Technology use, self-directed learning, student engagement and academic performance: Examining the interrelations. *Computers in Human Behavior, 63*, 604–612.

Redmond, P., Heffernan, A., Abawi, L., Brown, A., & Henderson, R. (2018). An online engagement framework for higher education. *Online Learning, 22*(1), 183–204. https://doi.org/10.24059/olj.v22i1.1175

Reece, A. J., & Butler, M. B. (2017). Virtually the same: A comparison of STEM students' content knowledge, course performance, and motivation to learn in virtual and face-to-face introductory biology laboratories. *Journal of College Science Teaching, 46*(3), 83–89.

Reeve, J. (2013). How students create motivationally supportive learning environments for themselves: The concept of agentic engagement. *Journal of Educational Psychology, 105*(3), 579–595. https://doi.org/10.1037/a0032690

Revere, L., & Kovach, J.V. (2011). Online technologies for engaged learning a meaningful synthesis for educators. *Quarterly Review of Distance Education, 12*, 113–124.

Reychav, I., & McHaney, R. (2017). The relationship between gender and mobile technology use in collaborative learning settings: An empirical investigation. *Computers and Education, 113*, 61–74.

Ryan, R. M., & Deci, E. L. (2020). Intrinsic and extrinsic motivation from a self-determination theory perspective. Definitions, theory, practices, and future directions. *Contemporary Educational Psychology, 61*, 10–18. https://doi.org/10.1016/j.cedpsych.2020.101860

Shailaja, J., & Sridaran, R. (2014). Taxonomy of e-learning challenges and insight to blended learning. *International Conference on Intelligent Computing Applications, 14*, 310–331. https://doi.org/10.1109/ICICA.2014.70

Singh, D., & Mishra, S. (2019). Benefits and effect of e-learning resources on the academic performance of the student. *International Journal of Research and Analytical Reviews (IJRAR), 6*(2), 533.

Stone, C. (2016). *Opportunity through Online Learning: Improving Student Access, Participation, and Success in Higher Education.* National Centre for Student Equity in Higher Education.

Stone, C. (2019). Online learning in Australian higher education: Opportunities, challenges, and transformations. *Student Success, 1*(2), 1–11. https://doi.org/10.5204/ssj.v10i2.1299

Thomas, J., & Graham, C. (2019). Online teaching competencies in observational rubrics: What are institutions evaluating? *Distance Education, 40*(1), 114–132.

Vella-Brodrick, D., & Klein, B. (2010). Positive psychology and the internet: A mental health opportunity. *E-Journal of Applied Psychology, 6*(2), 30–41.

Weil, S., McGuigan, N., & Kern, T. (2011). The usage of an online discussion forum for the facilitation of case-based learning in an intermediate accounting course: A New Zealand case. *Open Learning: The Journal of Open, Distance and e-Learning, 26*, 237–251.

Welsh, E. T., Wanberg, C. R., Brown, K. G., & Simmering, M. J. (2003). E-learning: emerging uses, empirical results, and future directions. *International Journal of Training and Development, 7*(4), 245–258.

Woods, K., & Bliss, K. (2016). Facilitating successful online discussions. *The Journal of Effective Teaching, 16*(2), 76–92.

Wu, W. C. V., Hsieh, J. S. C., & Yang, J. C. (2017). Creating an online learning community in a flipped classroom to enhance EFL learners' oral proficiency. *Journal of Educational Technology & Society, 20*, 142–157.

Yeld, N. (2016). Can UN development goals fix higher education problems? Retrieved from https://www.britishcouncil.org/voices-magazine/can-un-development-goals-fix-higher-educationsproblems

Zainuddin, Z., & Halili, S.H. (2016). Flipped classroom research and trends from different fields of study. *International Review of Research in Open and Distributed Learning, 17*, 313–340.

Zare, M., Sarikhani, R., Salari, M., & Mansouri, V., (2016). The impact of e-learning on university students' academic achievement and creativity. *Journal of Technical Education and Training, 8*(1), 25–33.

Chapter 8

Contextual Learning

P. Janardhana Kumar Reddy and K. Revathy

Learning Objectives

- To understand and define the concept of contextual learning.
- To understand the theoretical foundations of contextual learning.
- To appreciate the multiple compartments of the real world in contextual learning.
- To understand the role of various contexts and the process of contextual learning.
- To be able to identify the various components and characteristics of contextual learning.
- To be able to implement contextual learning using various techniques.
- To understand the role of technology as a driver of contextual learning.
- To distinguish between contextual and traditional learning.
- To appreciate the significant benefits of practicing contextual learning.

Introduction

Learning is a type of cooperation between the individual and the climate that offers the experience of the genuine circumstance. This sort of connection prompts a progression of learning encounters. In this view, Burton et al. (2006) states "a decent learning circumstance should comprise of a rich and changed grouping of learning encounters around a powerful incorporated reason and be carried on in interface with a rich, different and invigorating climate". Learning shows the association among students, instructors, and learning assets in a learning climate. Learning

DOI: 10.1201/9781003504894-8

inside an instructive unit must be coordinated in an intuitive, spurring, invigorating, and testing way, in this way uplifting understudies' dynamic cooperation. Learning gives students the possibility to foster drive, imagination, and autonomy as indicated by their abilities, interests, and physical and mental turn of events.

Today, it is generally acknowledged by instructors and scientists that learning ends up being more successful and significant when it happens inside a fitting setting that shows genuine characteristics.

Contextual Learning

Contextual learning might be characterized as a constructivist strategy for guidance that works with understudies to apply new information and abilities to genuine circumstances. Relevant learning is true outside-study hall experience inside a specific setting which fills in as an impetus for students to use disciplinary information for their own and expert turn of events. Past the test of direct huge encounters, context-oriented learning expects reflection to make dependable intellectual associations. This methodology includes understudies' dynamic interest in the learning system to make and revive their insight and encounters logically. The instructor presents data to such an extent that the understudies can develop importance based on their own experiences.

Key Concepts of Contextual Learning

Contextual learning enables students to process new information provided to them in ways that add up to their sense in terms of their world of memory, experience, and response (Parnell, 1995). The core concept behind contextual learning includes the following features:

- Knowledge is fabricated through individual/social activity, reflection, and interaction.
- Student-centered, authentic, self-regulated learning with the teacher as a facilitator.
- Application of newly acquired knowledge and skills to real-life situations.
- Constructing one own memory, experience, and therefore the response of the learner.
- Authentic and real-life opportunities for application.
- Having the capability to make use of education in a variety of contexts both inside and outside of the classroom.
- Rich set of fun-filled learning experiences in different contexts.
- Contributes to meaningful, relevant, and useful learning.
- Motivation is the key to learning.

Theories Behind Contextual Learning

Hull's Contextual Learning Theory

Hull's Contextual Learning Theory states that learning happens only when the learners process new information or newly acquired knowledge in ways that make sense in their frame of reference concerning their own experience, memory, and response. Therefore, a particularly close-to-home reference outline is generally sustained by the singular's collaborations and encounters with this present reality (Figure 8.1).

This methodology accepts that the brain normally searches for context-oriented importance and it does so by inspecting connections that bode well and look valuable. As indicated by this technique for guidance, the cerebrum chases for the relevant significance of specific connections inside our learning surroundings (Johnson, 2002; Powers and Guan, 2000). Based on this thought, the logical learning hypothesis centers around the different parts of a learning environment like study hall, lab, or work environment. It likewise urges teachers to pick or configure learning conditions that coordinate various types of learning encounters for accomplishing wanted learning results. In such a climate, the understudies learn significant connections between theoretical thoughts and viable applications in true settings of this present reality; ideas are disguised through the course of disclosure, support, and relating.

Experiential Learning Theory

In the mid-1900s, John Dewey (1938) fostered his hypothesis of experiential discovery that focused on the significance of having legitimate encounters. He proposed

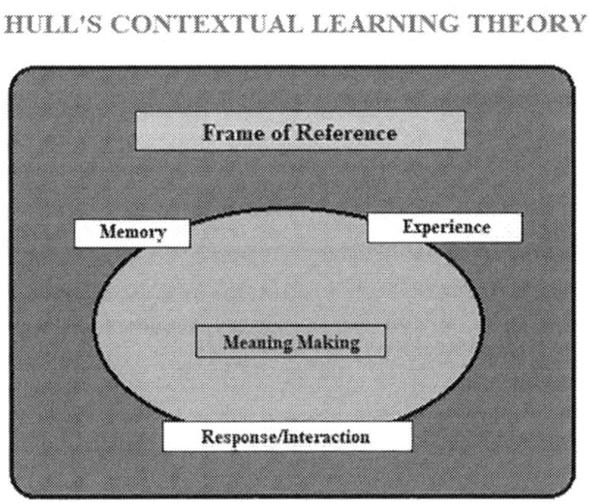

Figure 8.1 Hull's contextual learning theory.

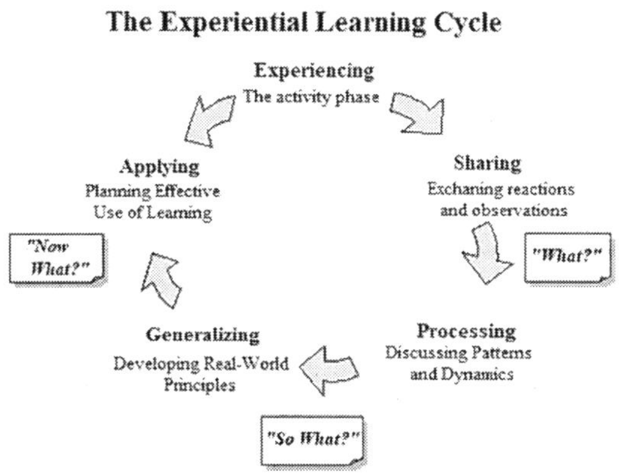

Figure 8.2 Experiential learning theory.

that learning ought to be contextualized and changed by genuine circumstances. He likewise centered on recognizing the interrelation, everything being equal (i.e., ideas), and proposed to focus on these linkages as opposed to on the actual things. This aim mirrors the insightful mix of hypothesis and practice, content and setting, of reasoning and activity. Dewey was excited about contextualizing learning content through request and critical thinking measures (Innes, 2004; Figure 8.2).

Cognitive Learning Theory

Cognitive learning theory explains how the human mind works as one's learning happens. It concentrates on the thought and mental processes and also on how they are influenced by both internal and external factors that construct learning within an individual. The capability of the brain's mental processes to take in and retain information in terms of experience, senses, and thought is known as cognition. This theory places key importance on cognition as part of the learning process. Like a computer, the human mind absorbs information, processes it, and then uses it to produce learning outcomes (Figure 8.3).

This theory is largely based on the studies of Jean Piaget (1966) who rejected that learners are passive, simply responding to environmental stimuli. As per this theory, the acquisition of knowledge is not passive but is achieved through active construction by cognitive structures through regular practice. According to Bruner (1977), discovery learning can support students to better understand and learn concepts and principles through active participation. Besides, Carin and Bass (2001) express that instructing must be considered as a request-based methodology that utilizes logical abilities to direct understudies toward revelation. The request approach puts accentuation on learning through student-focused encounters. Understudies ought to be offered the chance to look at a marvel utilizing the idea and the related information to accomplish an end.

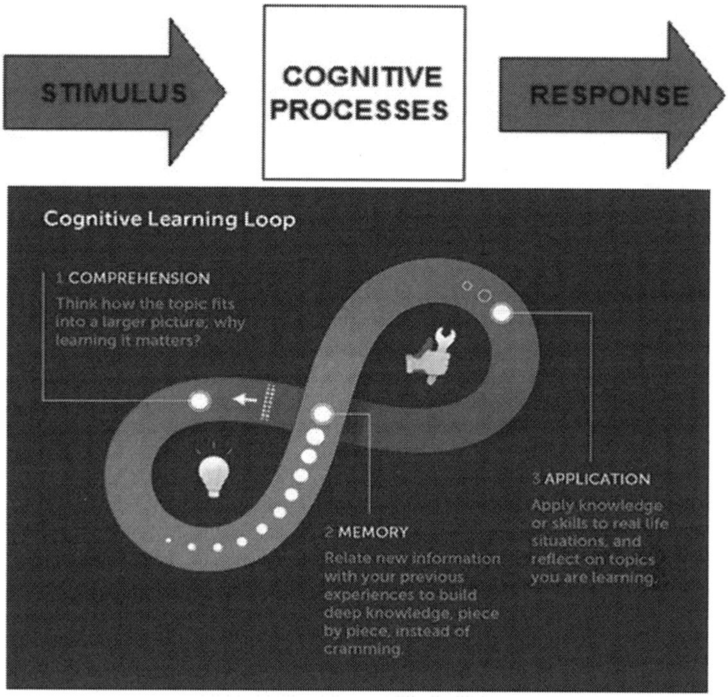

Figure 8.3 Cognitive learning theory.

Connectivism Theory

From the perspectives of Berns and Erickson (2001), "relevant learning empowers understudies to associate the substance that they figure out how to the existence settings where the learned substance could be utilized." Learning happens at the point of past information, experience, knowledge, reality, comprehension, and flexibility. Learning is dispersed inside an interpersonal organization with mechanically improved, unmistakable, and interpretable examples. Learning is acknowledged through Collaborating, Constructing, and Intensifying through friendly and significant associations. The connective hypothesis joins the impacts of three distinct instruments of the mayhem hypothesis, organizations, and the exchange of intricacy and self-association.

- **Chaos Theory:** The thought behind Chaos Theory is that, regardless of how random occasions could appear, when concentrated together, they make an example that could show significance past the singular occasions themselves.
- **Networks:** Organizations are associations among the elements related to the learning measures. Through these organizations – of people, of innovation, of social designs, of frameworks, etc., the students can impart their plans to other people, in this way mixing their learning climate.

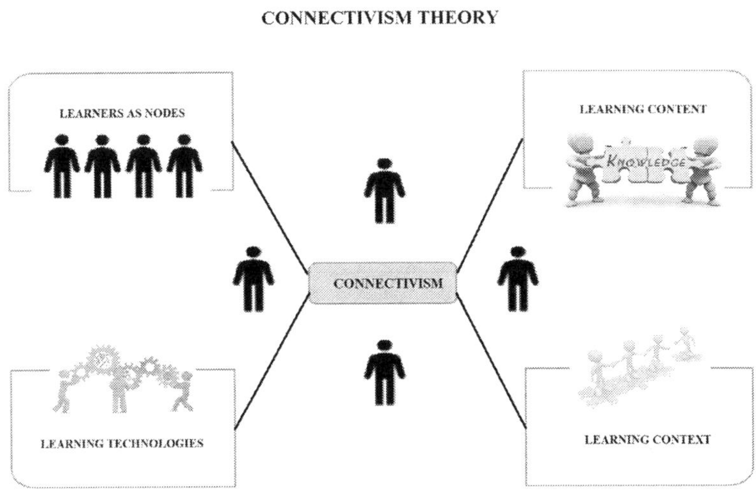

Figure 8.4 Connectivism theory.

- **Complexity and Self-Organization:** An intricate framework is a pool of connecting specialists that address parts that are just about as various as individuals, cells, or atoms. Inferable from the non-linearity of the cooperation, the development of the general framework is arbitrary and wild to some even out. However, the framework becomes self-arranging, because the singular associations eventually make worldwide synchronization and cooperative energy (Figure 8.4).

Constructivism

Constructivism centers on the development of information through the cooperation of existing information and recently gained information through experiential learning (Von Glasserfeld, 1992). It depends on the possibility that individuals assemble their insight through dynamic development and that students' encounters decide reality. Students utilize their earlier information as a base and assemble it with new learnings. Consequently, the learning is exceptional for each person as their encounters contrast with one another. This learning hypothesis insists that information is best acquired through a course of activity, reflection, and development. Every student adapts separately and socially to make significance to their experiential learning through a course of reflection and dynamic development in the brain. The constructivist structure forces the students to assume liability for their securing of information and urges the teacher to fill in as a facilitator. This model was established in learning hypotheses created by Dewey, Piaget, Vygotsky, Gagne, and Bruner (Figure 8.5).

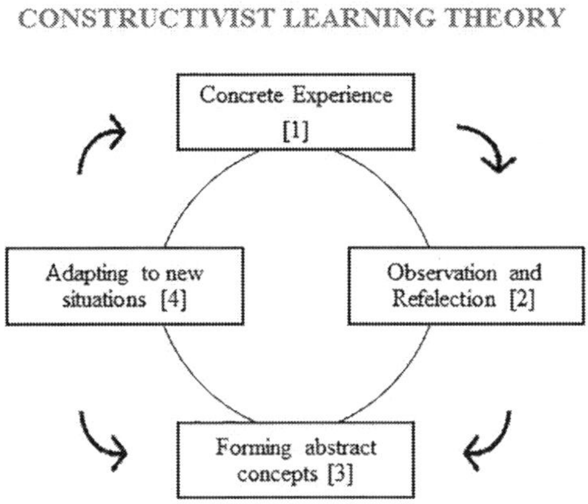

Figure 8.5 Constructivism.

Gardner's Theory of Multiple Intelligence

Gardner's theory portrays the idea of various knowledge. As per this hypothesis, various knowledge goes from the use of words, numerals, pictures, and music, to social communications, self-examination, reflection, sensation development, and being in line with nature. Instructors need to change their showing styles depending on the understudies' insight types. Pluralization recommends that themes and abilities ought to be instructed in more than one way, consequently initiating a person's numerous insights. Introducing a spread of exercises and ways to deal with learning urges all understudies to have the option to ponder the taking in content according to different viewpoints, subsequently extending their insight into the subject instructed (Figure 8.6).

Dewey's Reflective Learning Theory

John Dewey's reflective learning theory significantly links to personal learning. Dewey (1933) highlighted that reflection within a learning context is not just a passive recall of an experience/event. Reflection is a thoughtful and dynamic process. It is about rationality to learn. Dewey thought of reflection as an obsession with things that puzzle or disturb us and saw reflection as a sort of antecedent to action. He supposed that reflection could be truly beneficial for building meaning out of puzzling situations or events that are hard to elucidate. He advocated that reflective learning embraces recalling the event and then probing inquiries to explore why things happened the way they did and what possible actions could have given a different upshot. Since Dewey's initial work, many models have been proposed to recognize the components of reflection within learning (Figure 8.7).

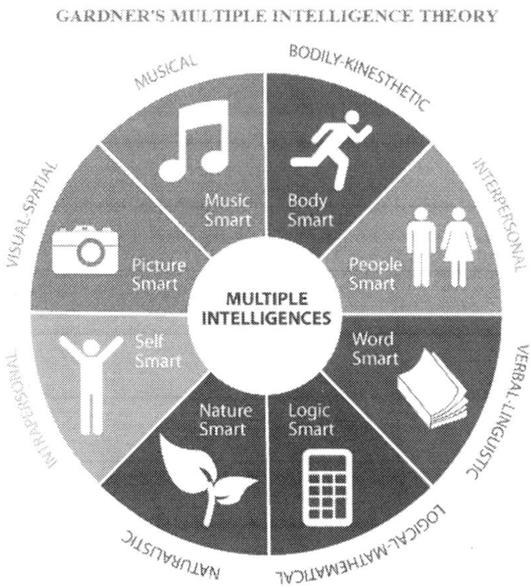

Figure 8.6 Gardner's theory of multiple intelligence.

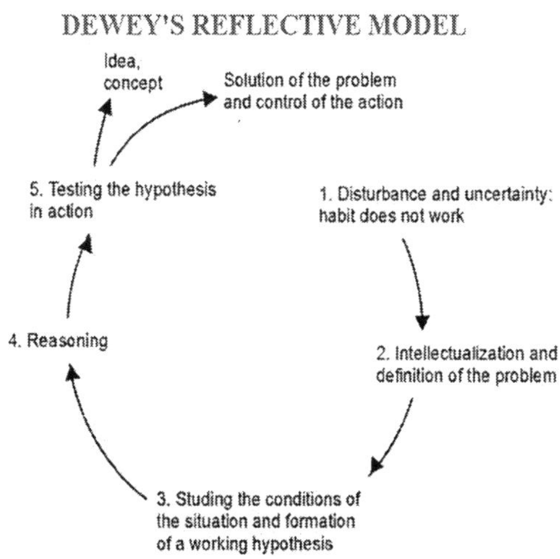

Figure 8.7 Dewey's reflective learning theory.

Situated Learning Theory (SLT)

Situated learning is an educational approach developed by Jean Lave and Etienne Wenger in 1991, to explain the process and development of learning when individuals get the opportunity to participate in a practice community.

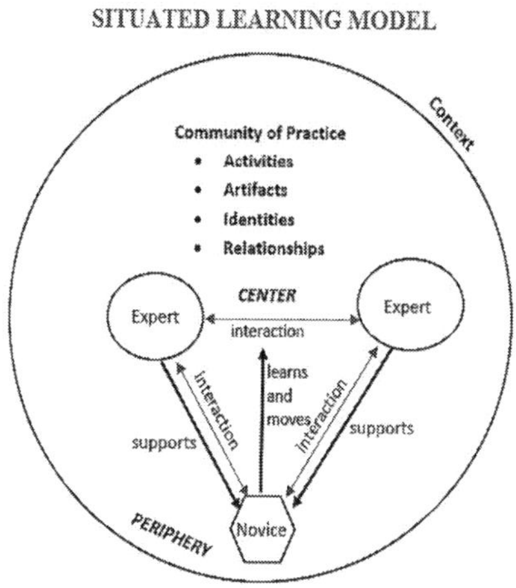

Figure 8.8 Situated learning theory.

In such a community, new learners attain the levels of an expert as they gain more practice opportunities within the learning context. It trails the work of Dewey, Vygotsky, and others who explained that students are more inclined to learn through active participation in the learning experience (Figure 8.8).

Situated learning fundamentally is a method of making meaning from day-to-day real-life activities (Stein, 1998) where learning happens relevant to the learning environment. The following concepts serve to define SLT

1. **Learning**, the skills, and knowledge attained as a result of participation in a practice community;
2. **Context**, the setting within which the situated aspect of learning happens;
3. **Novice**, the learner and newcomer in the practice community who learns from the experts around; and
4. **Experts** are the group of individuals who have the expertise in the subject to be learned.

Kolb's Model of Learning

Kolb (1984) proposed an interaction-based methodology focused on a superior adjusting among hypothesis and practice. Kolb's key concern was to diminish the abundance accentuation on the solidness of this present reality practice by broadening substantial learning encounters with the foundation of summed-up mental models and theories. Kolb's learning model clarifies that learning spins around the procurement of dynamic ideas and that they can be applied adaptably in a scope of circumstances. In Kolb's hypothesis, the inspiration for the improvement of new ideas is upgraded by new experiences. David Kolb's review shows that students

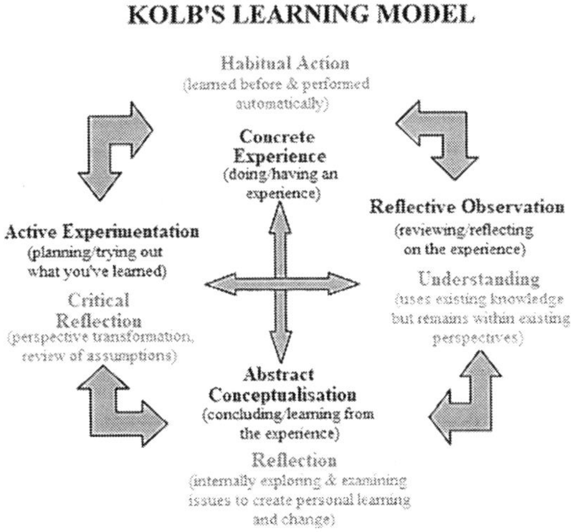

Figure 8.9 Kolb's model of learning.

see data either uniquely (by conceptualizing/thinking) or solidly (by encountering/ feeling) and afterward measure data either effectively (by testing/doing) or brilliantly (by noticing/watching) (Kolb and Kolb, 2005). Effective learning is seen when an individual advances through a pattern of four phases:

1. Gaining a concrete experience tracked by
2. Observation and reflection of that experience leading to
3. Abstract Concepts Formation (analysis) and generalizations (conclusions) used to
4. Test a hypothesis in prospective situations, causing new experiences (Figure 8.9).

The 360° of Context

Context-oriented Learning (CL) can be characterized as a learning framework that serves to interface cerebrum cycles to make significant examples by associating scholarly substance with the setting of reality. This assists with putting away them not just in transient memory, which as a rule gets effectively forgotten by the understudies, yet in addition in long haul memory which would assist them with applying these in their day-by-day schedules and working environment commitments at later stages in their day to day existence.

This is a context-oriented methodology as it assists instructors to interface scholastic substance with certifiable circumstances that the understudies insight. This rouses understudies to fabricate associations between their obtained information

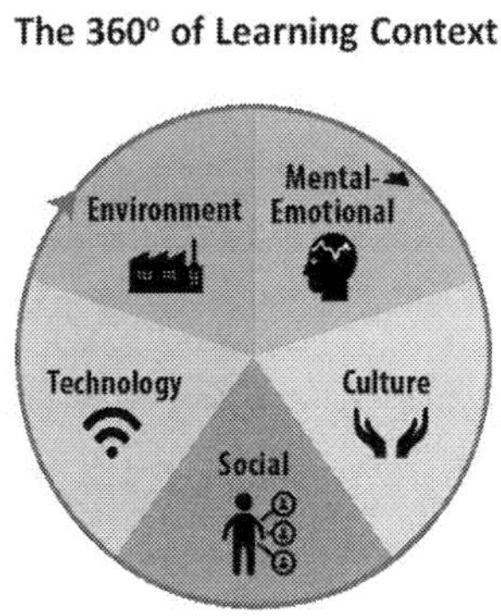

Figure 8.10 The 360° of context.

and their lives. Setting means "a staggered work of components that installs learning and execution" (Tessmer & Richey, 1997). Setting empowers students to straightforwardly connect ideas with their true reciprocals and set their gained information in motion. The setting of learning can incorporate the accompanying as displayed in Figure 8.10:

- Physical Environment
- Mental-Emotional State
- Culture
- Social Situations and Relationships
- Technology

Context arises from the interactions of an individual with the outside world as given in Figure 8.11. The different compartments of the world can be differentiated as follows:

- The **individual learner** the singular student lies at the focal point of the figure and is encircled by various causal regions of the planet; the collaborations of the student with these lead to **the learning context**.
- The strongest compartment for the singular student is the substantial functional setting where the individual is acting. It is the world that we straightforwardly see and follow up on. It reflects the "here" and "presently" of one's being, and could allude to a specific area, a structure or a room, and the close by items, and individuals.
- The compartment of space information indicates the topic occupied by the students. This incorporates a subsection of human information like language, mathematics, history, design, or cooking. Every one of these spaces has its jargon, techniques, and instruments, hence making it a unique circumstance.

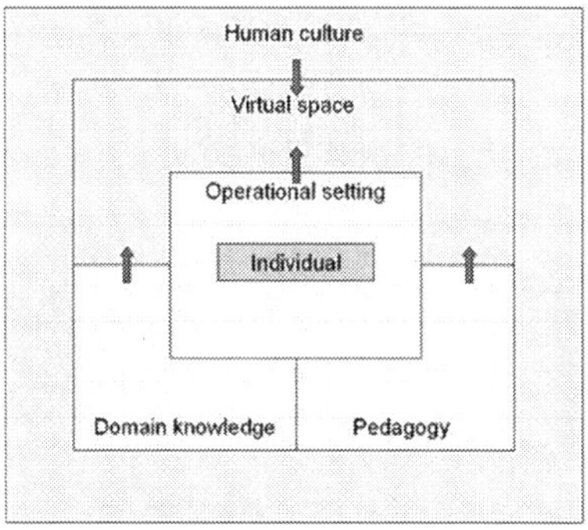

Figure 8.11 Compartments of the world.

- The compartment of the instructional method examines the assorted learning and instructing systems that characterize the various jobs and obligations of students and educators, the learning exercises, and the manners by which direction, input, and testing are organized. For instance, homeroom learning includes an alternate setting of realizing when contrasted with a discipline at the organization (Fuller et al., 2004).
- The external shell portrays the world everywhere, specifically human culture. It is the sweeping and interrelated entire of thoughts, information, convictions, expressions, laws, ethics, abilities, and customs merged across fluctuating ages and networks (Cassirer, 2006).
- Virtual space alludes to the advanced augmentations of the world open through computerized gadgets in the functional setting. It offers advanced assets and instruments, permitting correspondences with the rest of the world. It practically spreads the human skyline of cooperation past the actual limits of the functional setting. The bolts in the figure above show the propensity of expanding virtualization of the world; the virtual compartment subsequently acclimatizes substances that start from different compartments, notwithstanding, it likewise goes about as a divert in its particular manner.

The "REACT" Strategy

The REACT strategy of related learning is built on the core constructivism ideologies. Contextual learning, according to Hull and Greveelk (1998), has five significant features, which include connecting, experience, applying, collaborating, and displacing/transferring. Contextual Learning curriculum and instructions should be organized to inspire these five essential characteristics of learning. The REACT strategy focuses on Relating, Experiencing, Applying, Cooperation, and Transfer (REACT) in a linear fashion in the process of contextual learning (Figure 8.12).

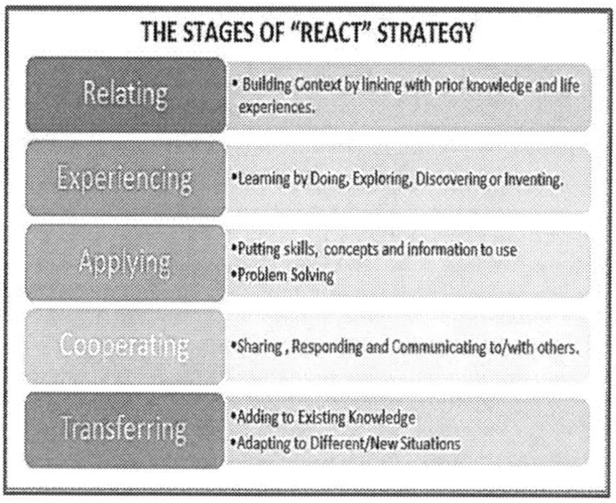

Figure 8.12 The REACT strategy of contextual learning.

Relating

Experiencing refers to learning in the context of exploration, discovery, and invention. Learning happens faster when students use equipment and materials to do active research by themselves. It enables them to practice actions connected to real life. Experiential learning can also be encouraged through multimedia and ICT-enabled technologies. It also helps in retaining the learned content according to the intensity of the student's learning experiences.

1. **Experiencing**

   Experiencing refers to learning about investigation, revelation, and innovation. Learning happens quicker when understudies use gear and materials to do dynamic examinations without help from anyone else. It empowers them to rehearse activities associated with reality. Experiential learning can likewise be supported through media and ICT-empowered advancements. It likewise helps in holding the learned substance as indicated by the power of the understudies' learning encounters.
2. **Applying**

   Applying refers to employing/putting the learned concepts and information in a useful context. Teachers can motivate students by making problems more realistic and relevant to students' lives. This stage helps in carrying out the concepts or skills that are learned by the students into real-life scenarios and solving problems. Learning is heightened when concepts are presented in the context of their usage.
3. **Cooperating**

   Cooperating refers to learning about sharing, reacting, and teaming up with different students. Working in bunches assists understudies with tackling complex issues through communitarian learning. This training assists understudies with comprehension becoming familiar with the idea and associating the learned substance with this present reality. This experience would help them in their workplace at later stages to impart viably, share data suitably, and work serenely in a group.

4. **Transferring**

   Transferring refers to learning in the context of existing or newly acquired knowledge in a new context or situation. It uses students' prior knowledge and builds on it.

   This differs from the "relating" stage as it does not involve new learning but, focuses on building a new learning experience on what one already knows. In this stage, the teacher makes students extract what they have learned and apply it to new situations and contexts.

The Process in Contextual Learning

Contextual Learning uses brain research to demonstrate the significant role of meaning and connections in learning a specific lesson. It is an idea that helps instructors to associate topics with genuine circumstances. It urges students to represent their learning and to make linkages among information and its applications to the different life settings of family, companions, work environment, etc. This approach considers content as an important component and involves students' active learning process. This approach to learning can be implemented individually or in groups. The technique pays care to the two aspects: involvement of the student's active participation and relating their knowledge within the contexts of society, culture, and daily life. The students evolve with more competence in the learning content as they start using them contextually in real-life situations (Figure 8.13).

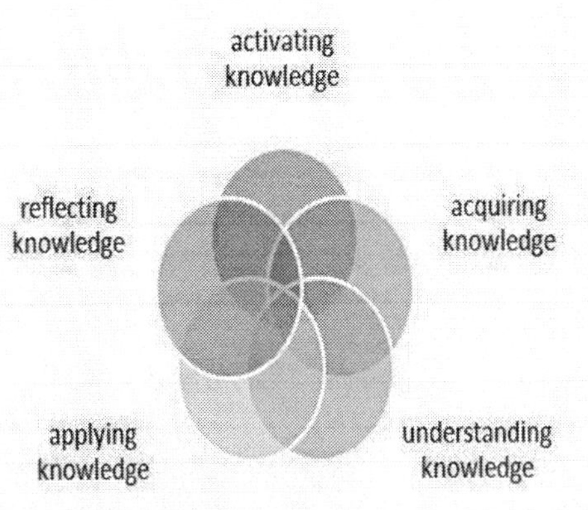

Figure 8.13 Contextual learning process.

Contextual Learning emphasizes the brain's processes to capture and retain knowledge by establishing meaning and connections. This process involves the following mental processes:

- Acquiring Knowledge
- Understanding Knowledge
- Applying Knowledge
- Reflecting Knowledge
- Activating Knowledge

The brain processes information by way of learning contextually through the coordination of the above-mentioned cognitive processes.

Contextual Learning includes the following three key features in the learning process:

Situated Cognition -
- Applied Knowledge in multiple real-life contexts.

Social Cognition -
- Intrapersonal Constructs in connection with other people.
- Distributed Cognition - Constructs continually shaped by other individuals and things outside the individual.

Characteristics of Contextual Learning

Contextual Learning is an instructional strategy that includes the following framework characteristics:

- Learning focuses on the active involvement of the students in the learning process.
- Students view the learning as relevant.
- Learning combines content and context; all learning is applied knowledge.
- Students learn from each other through collaboration, interaction, group work, and self-reflection.
- Learning employs authentic materials and authentic assessment.
- Learning is linked to real-world and/or simulated issues and meaningful problems.
- Learning is done in an authentic context to emphasize higher-order thinking and develop problem-solving skills and competencies in the students.
- Learning uses openings for understudies to do significant tasks.
- Learning is done by giving a large number of significant encounters to understudies.
- Learning is accomplished through a bunch of work, conversations, and common rectification.

- Togetherness, participation, and understanding each other from top to bottom are pleasant parts of learning.
- Learning is accomplished through dynamic, inventive, useful, and helpful exercises.
- Students are urged to assume liability for the observing and advancement of their adapting in this manner advancing self-controlled learning.
- Students are spurred to be dynamic members in the improvement of society.
- Student learning is assessed by numerous strategies.
- The perspectives and contemplations of understudies are esteemed and regarded.
- The teacher assumes the part of a facilitator of understudy learning.
- The teacher acknowledges his/her job as a student. The teacher employs diverse teaching techniques and approaches.
- The learning environment is dynamic and exciting to promote joyful learning.
- Students and educators are invigorated to try different things with new methodologies advancing inventiveness.
- The interaction of learning is given due significance comparable to the setting that is learned.
- Valuing understudies' different life settings and past encounters is essential to learning.
- Learning happens in different settings and permits understudies to distinguish and tackle issues in new settings.
- Knowledge is considered as an interdisciplinary reaching out past the limits of regular homerooms.

Implementation of Contextual Learning

As per the perspectives of Berns and Erickson (2001), there are five techniques for executing relevant learning. These incorporate the accompanying

- The first is issue-based learning. It is an instructive technique where complex certifiable issues are utilized to advance understudy learning of ideas and standards as differentiating to the immediate show of realities and ideas. This methodology empowers the understudies to tackle the given issue as they are acquiring by coordinating abilities and ideas from many substance regions. It can start with either a genuine or reenacted issue. It assists with improving the improvement of basic reasoning, critical thinking, and relational abilities. It can likewise offer freedom to work in gatherings, finding and assessing research materials, and deep-rooted learning.
- The second is helpful learning. It is an informative technique wherein little gatherings, each having understudies of various capacity levels, utilize an assortment of learning exercises to work on their comprehension of a subject.

It intends to sort out homeroom exercises into scholastic and social learning encounters for understudies in little gatherings, each chipping away at a typical task. It gives them a chance to understudies to associate with one another and empowers them to acquire important social abilities.

- The third is project-based learning. It is an informative technique, wherein understudies achieve information and abilities by exploring and reacting to a bona fide, drawing in, and complex inquiry, or challenge inside the specified time-frame. It focuses on the guideline of discipline in directing the examination to take care of the issue and offers the chance for understudies to work independently.
- The fourth is administration learning. It is a sort of experiential learning strategy where understudies apply their scholastic information, basic reasoning, and critical thinking abilities to address real local area needs. It gives the genuine practice to foster information through project works and gathering exercises. It pushes the understudies past the constraints of the study hall to empower them as dynamic members of the learning system and to foster local area information and abilities.
- The fifth is **work-based learning.** It is workplace context-based learning which is required for the students for their job performance during their employment. It is the instructional strategy that provides students with real-life work experiences that can help them apply their academic and technical skills thereby developing their employability. It includes activities such as internships, workshops, mentoring, etc.

In addition to these, the following approaches can also be employed for implementing contextual learning inside or outside of the classroom.

- **Situated learning** - Situated learning intends to have an idea and activity which is utilized at the ideal opportunity, and spot. In this methodology, the substance is learned through doing exercises and is secured inside the setting.
- **Case-based learning** - This method is an instructional approach focusing on the intense interactions between the learners and the participants as the learners build their knowledge base by working together in a group to examine the case under study. The instructor acts as a facilitator whereas the students team up collaboratively to analyze and resolve problem questions that do not have a single right answer.
- **Inquiry-based learning** - Inquiry-based learning is an active learning method that involves posing questions, problems, or scenarios that trigger curiosity. Inquiry signifies active involvement through probing questions for better understanding.
- **Reflective learning** - Reflective learning requires students to think about what they have done, or learned linking to their own lives, and making meaning

out of the learned material. It allows for the recall of already lived experiences aimed at developing critical thinking and improving future actions based on the analysis of the lived experiences.

The classroom practice for contextual learning may include the following features:

- Connecting content to student encounters.
- Engaging understudies in dynamic learning.
- Enabling understudies to have a few chances to coordinate their learning both exclusively and cooperatively.
- Encouraging the development of emotional significance from individual and aggregate encounters.
- Enabling the understudies to incorporate their gained information and abilities inside suitable settings of their genuine world as they learn.
- Assessing the learning results inside a credible circumstance alongside taking into account the cognizance of different implications out of a solitary encounter.
- Identifying content that is formatively fitting to the student.

Technology as a Driver of Contextual Learning

Contexts for learning can be regularly settled in various ways. Advanced media will in general reason new components of setting. It is critical to decide the appeal of this carefully upgraded setting and its significance for learning. Different organization and media advances acquire that learning isn't any more limited to fixed areas like schools, yet is regularly broadened to join various settings while supporting the work environment and getting the hang of gaining from home, area-based learning, or learning to progress. Students have unrestricted admittance to any information asset, banters in conversation sheets, contextual analyses, effective reports, certifiable video accounts, and so forth Immovably dependent on constructivism, investigation-based learning, and request-based learning have acquired energy among instructors.

Web 2.0 advancements everywhere will in general rethink the strategy chain of content creation while these empower students to make, share, and adjust their substance and assess these in online media organizations of companions, friends, partners, or others that don't share the indistinguishable illustration or homeroom. New data and specialized devices like cell phones, geo-situating administrations, surrounding conditions, and universal access expand the student's actual scope of activities by empowering increased reality layers superimposed on winning settings. Sensors, additionally as following and following innovations, give the contributions to setting elements through versatile frameworks conduct and personalization. In light of these turns of events, the learning setting turns out to be more powerful and responsive yet in addition incredibly immaterial and wild.

Comparison of Contextual and Traditional Instructions

Table 8.1 Comparison of Contextual and Traditional Instructions

S.No.	*Aspects*	*Contextual Learning*	*Traditional Learning*
1.	Approach	Student-Centric Approach	Teacher-Centric Approach
2.	Student Participation	Active engagement in the learning process	Passive recipients of information
3.	Learning Aspect	Concrete and Practical Learning	Abstract and Theoretical Learning
4.	Relevance	Linked to real-life or simulated problems.	May or may not link to real-life problems.
5.	Behavior	Built on self-awareness	Built on habits
6.	Socialization	Students learn from peers through group work, discussion, and mutual correction.	Students do individual learning.
7.	Learning Place	Learning occurs in a variety of places, contexts and settings. It can include both inside and outside of the classroom.	Learning only occurs in the classroom.
8.	Skill Development	Based on Understanding	Based on Training
9.	Good Behavior	Based on intrinsic motivation leading to self-satisfaction.	Based on Extrinsic motivation leading to praise or value.
10.	Learning Outcomes Measurement	Authentic assessments focused on emotional regulation, problem solving, teamwork, individual behaviors, and relationship management and so on.	Paper-Pencil/Pen tests, and academic exams

Significance of Contextual Learning

Contextual learning is a deliberate instructional practice that focuses on connecting the theoretical concepts taught in the learning process, to the practical and real-world context. By implanting instructions in contexts familiar to the learners, they could readily understand and integrate those instructions. This realization is firmly based on constructionist learning theory, which states that the learners learn better when encouraged to construct linkage between the instructions they receive, and their interpretations within the context of their environments.

Contextual learning approaches have been proven to be strongly rooted in Pedagogical Theories. Contextual learning activities are associated with the conventional pedagogical body of knowledge that includes theories of Motivation, Social Learning, Problem-centered Learning, and present psychological and physiological research that revolves around how human brains learn. In addition, it includes the following benefits:

- It highlights real-world application and problem-solving through higher-order thinking.
- It presents teaching through students' real-life experiences.
- It comprehends the need for instructing and learning in different settings - home, local area, and work environment.
- It encourages understudies to be responsible for their own adapting subsequently assisting them with becoming autonomous students.
- It assists understudies with gaining from one another and together in bunches sharing information and exercises.
- Contextual learning can help bring relevance, meaning, and application to the learning, thereby helping students to connect to the world they live in.
- Students can link the learned concepts in real-life situations and convert them into their life experiences.
- Increasing one's efforts results in more ability for healthy living.
- It motivates and encourages the learners to introspect themselves and reflect on their learning experiences thereby exploring their talent, skill, and learning levels.

Conclusion

Context-oriented learning is predicated on the idea that learning can't occur in a vacuum, and must in some manner be associated with confirmable perspectives to make sense to students. Such a setting empowers students to straightforwardly interface ideas with their genuine counterparts and set information right into it.

Integrating the ideologies of contextual learning in the instructional process helps the teachers to promote authentic learning within the students and hence the students can build their knowledge base by making connections to their already existing knowledge and the real-world settings, reflecting on their learning experiences, and applying their contextual learning in useful situations as they learn. This makes their learning more meaningful, relevant, useful, and authentic. The intensified wealth and intricacy of logical learning are incited by new computerized advancements. These new advancements get through the constraints of school structures and auditoriums by including broad and far and wide computerized assets and certifiable portrayals. The consistent progression of rising advanced media sticks to adding new measurements to the learning setting. Additionally, the reformist use of computerized media needs deliberate requests of the student's encounters, commitments, and necessities for these eminent learning settings.

References

Berns, R. & Erickson, P. R. (2001). *Contextual Teaching and Learning: Preparing Students for the New Economy.* Washington: National Academy Press. https://files.eric.ed.gov/fulltext/ED452376.pdf

Bruner, J. (1977). The Process of Education. Revised Edition. Harvard University Press.

Burton, Jeremy et al. (2006). Developing Conceptual Framework for Creativity, ICT and Teacher Education. Thinking Skills and Creativity, 1(1), 3–13.

Carin, A., & Bass, J. E. (2001).*Teaching Science as Inquiry.* 9th edition, Hoboken, NJ: Prentice-Hall, Inc.

Cassirer, E. (2006). *An Essay on Man. An Introduction to a Philosophy of Human Culture.* Felix Meiner Verlag eBooks. https://doi.org/10.28937/978-3-7873-2037-0

Dewey, J. (1933). *How We Think: A Restatement of the Relation of Reflective Thinking to the Educative Process.* Boston, MA: D.C. Heath & Co Publishers.

Dewey, J. (1938). *Experience and Education.* New York: Macmillan Company.

Fuller, A., Munro, A., & Rainbird, H. (2004). *Workplace Learning in Context.* Routledge.

Hull, D., & Greveelk, J. H. (1998). *Technical preparation: The Next Generation.* Waco Texas: Center Occupational Research and Development.

Innes, R. B. (2004). *Reconstructing Undergraduate Education: Using Learning Science to Design Effective Courses* (1st Ed.). Routledge eBooks. https://doi.org/10.4324/9781410610829

Johnson, E. B. (2002). *Contextual Teaching and Learning.* SAGE Publications.

Kolb, D. A. (1984). *Experiential learning: Experience as the Source of Learning and Development.* Englewood Cliffs, N.J. : Prentice-Hall.

Kolb, A. Y., & Kolb, D. A. (2005). Learning Styles and Learning Spaces: Enhancing Experiential Learning in Higher Education. *Academy of Management Learning & Education*, *4*(2), 193–212. https://doi.org/10.5465/amle.2005.17268566

Parnell, D. (1995). *Contextual Teaching Work Increasing Student Achievement.* Waco Texas: Center for Occupational Research and Development (CORD).

Piaget, J. (1966). The Psychology of Intelligence and Education. *Childhood Education*, *42*(9), 528. https://doi.org/10.1080/00094056.1966.10727991

Powers, S. M., & Guan, S. (2000). Examining the range of student needs in the design and development of a Web-Based course. In *IGI Global* eBooks. https://doi.org/10.4018/978-1-878289-59-9.ch013

Stein, D. (1998). *Situated Learning in Adult Education.* ERIC Digest No. 195.

Tessmer, M., & Richey, R. C. (1997). The role of context in learning and instructional design. *Educational Technology Research and Development, 45*(2), 85–115. https://doi.org/10.1007/bf02299526

Von Glasersfeld, E. (1992). Constructivism Reconstructed: A Reply to Suchting. *Science & Education, 1*(4), 379–384. https://doi.org/10.1007/bf00430964

Chapter 9

Experimental Learning – The Source of Development

S. Arulsamy and A. Benjamin

Introduction

> "For the things we have to learn before we can do them, we learn by doingthem."
>
> *– Aristotle, 350 B.C.E*

Every Educational Institution is a Learning Centre and promotes learning. Learning is a continuous process. All of us learn as students, teachers, parents, educators, and as humans. The form of promoting learning by each organization is a challenge in this competitive world. Every institution wants to give the best to its students and wants to see the students grow intellectually. As there is a shift from rote memorization skills to competence-based learning, it is the need of the hour for institutions to impart the best Teaching-Learning process. There is a demand in the country for testing students' higher-order skills such as analysis, critical thinking, and conceptual clarity to enter into various higher education courses. Therefore, every student should be given the best learning environment where they can validate the theory by performing experiments.

Learning should be experienced and it should be challenging enough to have problem-solving and decision-making ingredients in it. Experimental Learning is an important pedagogy in all educational institutions. To cope with and to enable our

DOI: 10.1201/9781003504894-9

students to acquire 21st-century skills, Experiential Learning has gained momentum. Every educator must imbibe the qualities and skills of the future in our learners, rather than acquiring the talent by simply rote memorization. Let us all come out of the four walls of the classroom, out of the limited pages of the textbooks, and let learning thrive and magnify in day-to-day life.

This chapter is useful for making the learners responsible for their learning. This is meant to change the way we teach, the way children learn, and the way skills are acquired. This essentially probes deeper into the essence of experimental learning and how to bring it to the class or at home to make learning joyful, engaging, and at the same time connected to real life.

Experimental Learning aims at making the educational environment student-centered. The students have control over their learning the pace of learning, the method of learning, and the skills they need to utilize for this learning. They can evaluate, think critically, make decisions, and master knowledge by constructing it. Learning experience may be cooperative, collaborative, or independent, encouraging the students to work together and learn how to question and evaluate evidence rather than accepting truths communicated by their teacher.

Objectives of Experimental Learning

- To place cognition and competence and utilize them in a meaningful manner
- To facilitate real-time coaching and learning for the students
- To accelerate the joy of learning from a novice to an expert
- To promote reflective thinking and face new opportunities and challenges
- To enable students to apply theoretical knowledge without any rigidity

Ways to Experience Experimental Learning

The following are the various ways to experience experimental Learning

1. **Hands-on learning/learning by doing:** Experimental learning aims at giving more importance to the students than the teachers as the method of learning shifts from a teacher-centric model to a student-centric model learning environment. Students are given the freedom to gain first-hand experience through real-life scenarios which sharpens their retention levels and helps accelerate learning. By protecting these skills in the classroom regularly, a student gets out of his/her comfort zone from simply being a listener in the class to a curious person who always wants to try out his/her theoretical knowledge can be put into practice. Thus, he/she can gain a wider perspective of the concepts being taught and it helps to bridge the learning gap between theory

and practice. Students must be encouraged to apply the concepts in real-life situations. A few examples are given below:

Magical Bubbles: Make this homemade bubble solution by placing the liquid in a narrow dish with a straw to blow bubbles. Let students explore how

Flare–up: Combine red/orange food coloring with two ounces of vinegar. Add two tablespoons of baking soda to a paper cup and place it inside a paper bowl. Quickly add vinegar to the cup using a cone. Wait to see the volcano erupt as children enjoy the flare–up! to blow and fill up the dish with as many bubbles as they can!

Sink or Sail: A fun activity to create a theory can be a great way of learning for the students. Various objects that sink or sail are tested in a tub filled with water. Toys and empty bowls with "Sink" and "Sail" written on them can also be checked.

2. **Talents are not sold in the market, but are there in every student:** "Education is the manifestation of the perfection already in man" (Swami Vivekananda). History has always proved that each individual is unique and different with many talents. Learning through the process of doing brings out the hidden potential of every student in the classroom. Students must be encouraged and motivated to share their ideas and creativity in the classroom. *Personalized learning* is highly demonstrated through experimental learning. The concept of the flipped classroom is used here, where the learning goes to the students.

   The holistic growth of every student must be the aim of an educational institution. A student's uniqueness can be tapped by giving opportunities that ignite a spark in the hearts of students.

3. **Out-of-the-box thinking:** To think from a new perspective is a desirable quality needed for the present student community. Since we are accustomed to being individuals of habit, we prefer to follow our normal routine. However, if we can show our students how to rearrange their minds by thinking in new directions, with their vivid imaginations, we can inculcate in them to become thinking students rather than mere learners.

   Every learner should be motivated by creativity and view every situation from a wider perspective. Out-of-the-box thinking generates ideas that come naturally when needs are looked at from fresh perspectives, with no guidelines

   Groundbreaker – You are disruptive, in a good way. Indifferent traits are correlated with pushing the frontiers of what is expected and what is possible.

   Inspiring – An indifferent style can provoke admiration, envy, and respect from those who live their lives more cautiously.

4. **Convert your study room into a research room:** Recreating an environment by converting the study room into a research room produces demonstrable changes in the mindset of the students. Experimental learning is personal by nature and influences a student's feelings as well as emotions. It also enhances

one's knowledge and skills and helps to learn at their own learning pace. As Rabindranath Tagore says "You cannot cross the sea merely by standing and staring at the water". From building models to making the best out of waste, a student's study room-cum-research room allows them to reflect and engage, and to focus their attention on the task at hand.

Home is the place where students learn to validate their theory with experimental learning with the help of basic instruments, like Lenses, Mirrors, Prism, and Maps. They will be self-directed to perform experiments.

5. **Connect the dots:** Learning is all about connecting dots. All the subjects are interrelated, hence connecting the learning experiences and experiments leads to the holistic growth and development of the student. Several teaching methods must be used to make learning more dynamic and interesting. Role plays, group discussions, quizzes, and multi-sensory learning are some of the most preferred approaches in the modern teaching-learning process. Every student must involve themselves and enjoy the process of learning.

   A simple saying "That's a brown chicken," can be replaced by stating that the chicken lays eggs and eats insects in the farmer's garden. Nurture the child's curiosity to see the possible causes and consequences. Draw the connections they see among cars, air, plants, and people.

   Simple cause-and-effect diagrams can help solve the underlying problems. A heated debate on insisting your son clean the room can be sorted by sitting down together and drawing out a plan to sort the issue. Here, both sides are caught in a closed causal loop working to bring required and effective outcomes (Sweeney, 2016).

6. **Nature as a teacher:** Nature is man's best teacher. A student can connect to real-life situations by communicating with nature. All five senses of learning - the sense of sight, hearing, taste, smell, and touch are utilized for this process of experimental learning. Incorporating nature tips can have a long-term impact on students' learning. Through nature, every student gets a real-life learning experience that enhances classroom learning and helps them to correspond to the surrounding circumstances. Students are naturally taught to appreciate and conserve nature for future generations. Showing a rainbow in the sky makes the students remember the colors of the spectrum in the order of increasing wavelength or decreasing frequency.

   Learning must move out of the four walls of the classroom by taking students on field trips to paddy fields, gardens, power plants, and biogas plants, helping the students to learn from Nature. Setting up various activities for different class levels with nature will enhance learning.

7. **Available resources as learning tools:** To facilitate experimental learning effectively and efficiently, making use of the resources that are immediately available can be used as an instructional tool for learning. These tools can be a supplement to classroom learning. This may involve the use of old refrigerators, fans, microwave oven, and other do-it-yourself activities

that teach concepts in a detailed manner and is an extension of classroom learning.

Students learn from all resources available to them and use them as tools for learning. For example, Concepts relating to physics are learned with the help of electrical equipment available at home like AC, Refrigerator, Oven, and Iron Box.

8. **Use of technological tools:** With the online mode of teaching termed as a "new normal" use of technological tools enhances teaching and learning. Tutorials, simulations, and other productivity tools like PowerPoint presentations, excel, spreadsheets, word processors are all used to empower students for higher-order learning. These learning platforms enable students to gain knowledge beyond the four walls of the classroom.

   Technological tools related to education include augmented Reality, Flipgrid, Video Games, Google Forms, and Virtual Reality which help students receive blended learning. Learning through technology offers new possibilities to develop the competencies needed for the 21st century. These interactive tools continually enhance the understanding and build new knowledge and skills for the students (Mcguire 2023).
9. **Art integration:** Learning through arts is a creative and innovative teaching-learning approach where art becomes the medium of learning key concepts of any subject of the curriculum. It achieves a two-fold objective: (i) it helps students engage and involve themselves in the learning process and (ii) it helps to gain a broader view of the hypothesis to create a connection through art form and another subject area to acquire a greater understanding in both.

   Art Integrated Learning (AIL) is a teaching-learning model where art gives an insight to understand concepts within any subject of the curriculum. Arts have the flexibility to accommodate age-appropriate opportunities for learners who can explore at their paces. This resonates with the experiential learning approach.

   From the stage of infancy, children are naturally curious and responsive. Graphic images and concrete skills such as exploring cereals or pulses in the kitchen and painting with fingers help children grow holistically. Mathematical concepts and imaginative stories help awaken and nurture their learning patterns. Children connect with feelings and innovation on a cognitive level to incorporate and simplify complex learning experiences.

   The preamble of the 'Seoul Agenda (2010)' emphasizes that

> ...arts education has an important role to play in the constructive transformation of educational systems that are struggling to meet the needs of learners in a rapidly changing world, characterised by remarkable advances in technology on the one hand, and intractable social and cultural injustices on the other.
>
> *(UNESCO, 2010, p. 1)*

Art promotes, supports, and implements learning and living to be a part of the natural process for all students including those with special needs and diverse cultures. A keen observation, as well as experimentation, makes learning concrete, creative, and rewarding for one and all.

Art has its language and breaks all barriers and boundaries. Integrating local arts and crafts in different school spaces like classrooms, corridors, different rooms, and outdoors leads to the holistic learning of students through multidisciplinary associations, freedom of expression and opportunity for introspection, one common platform for all, active involvement of every child, team co-ordination and group efforts, emotional expression and stability, appreciate diversity and respect for national and cultural heritage, Art Integration is an experimental life long journey where learning is a continuous process and is pervasive in all spheres of life.

Example of Application of Experience Experimental Learning in Various Subjects

The following are the examples of application of experience experimental learning in various subjects;

English as Primary Language

The language English, being a standard language for the mode of teaching must be taught perfectly so that students will be able to express themselves efficiently.

Hands-on activities like building a block with the words start by placing a high-frequency word on each block. The player removes a letter from the block and places it on top of the tower. Now, the other player has to make the right word before the next player takes his turn. This can be used for kindergarteners. For elementary and high school students, activities like preparing comic strips, writing a journal, chapter chats, dressing up like the characters can spark creativity and engage students in a love of literature.

Social Science

The subject of social science brings history back to life. Nature can be used as the domain for research to supplement classroom learning. Mapping our own home is a great way to learn about places around the world. Preparing travel brochures about India and stating the reason why India is one of the most sought–after destinations in the world can help students gain a broader understanding of our country. These activities also help students to develop a patriotic sense and love for the country.

Mathematics

Mathematics, being the subject that deals with shapes, quantity, and pattern is the building block for everything in our day-to-day life. Through experimental learning, mathematics can become a favorite subject for students. Activities like playing BINGO games, to learn basic concepts of addition, subtraction, multiplication, and division, preparing a survey, and presenting it through bar graphs and pie chats can help students build their skills to learn different sampling techniques. Real fruits are used for addition/ subtraction to make learning interesting in primary classes.

Accountancy

Accountancy is one of the key functions of every business and hence is a branch of commerce that deals with recording and summarizing every business transaction to help prepare the financial statements every year. Real-life examples of business are the best way to learn accounts. Through role-plays and group discussions, students are quickly able to see the connection between each concept. Creating a personal budget is a great tool to learn accounting. From the parents' income, a student learns to prepare an Income-Expenditure statement.

Business Studies

Business studies, being a broader subject in social science, deals mainly with management and entrepreneurship. A very interesting and lively subject, the use of current events, and newspaper articles are evaluated and discussion on how the situation could be resolved using managerial techniques is a great way to inculcate decision-making and critical thinking skills in students. The game "Deal or No Deal" is a unique activity where theoretical knowledge and the risk factors are explained for a business deal. The group then gives their final decision. Students get involved and engaged through this activity.

Economics

Economics studies deal with how people interact with the production, distribution, and consumption of goods and services. Trading games where students are directed into buyers and sellers by distributing cards of different colors agree on prices for commodities and mark their gain or loss. Auctioning is also another interesting activity where bidding takes place and concepts are learned with examples.

Activities demonstrated with experimental learning can build in students a variety of important skills like creativity, team bonding, critical thinking, and sportsmanship. Learning through active engagement and fun can leave a lasting impact and is considered to be a powerful instructional tool for students.

Experimental Learning – A Group Task

Teacher: Dear children, you are the best resource for doing an activity in school. In fact, this is a task for you all. Watch around you and list out a few resources that you can use for experiencing activities (Table 9.1).

Table 9.1 The Teacher Gives Them 10 Minutes and the Students Give the Following List

Readily Available Resource	*Possible Uses for Activities for Hands-On Experience*
Students	Role Plays, Games, Songs, Dances, Quiz, Art, etc.
School Playground	Games, Exhibition, Poster Competition, etc.
School Garden	Observing plants, species, soil, flowers, germination, etc.
Old Clothes, discarded objects	Waste Management
Computers	Hardware, Software
School Buses	Safety Drills, Transport Rules
Dance Room	Costumes, Culture orientation
Music Room	Instruments, Culture Orientation
Fee Counter	Basic Mathematics, Communication Skills
Principal's Office	Role Play, Communication Skills,
Water Taps/ Coolers	Pollution tests, Water Conservation, Recycling/Reusing
Mom's Kitchen	Knowing Spices, Seeds, Food Processing, etc.
News Papers	Reading skills, analyzing skills, evaluating skills, communication Skills, advertising, quiz, role play based on news items, etc.
Post Office, Local Administration Offices, Bus Stands, etc. (with a teacher)	Mode of Communication, Government Systems, Life Skills

> Children learn in a variety of ways—through experience, making and doing things, experimentation, reading, discussion, asking, listening, thinking and reflecting, and expressing oneself in speech, movement or writing—both individually and with others. They require opportunities of all these kinds in the course of their development.
>
> *(NCF, 2005 page 34)*

Experimental Learning Case Studies

Case I: "Curly Locks and The Three Bears" – Class II A

Let us first go to Mrs. Shakeera's Class II A. She is teaching English.

Mrs. Shakeera began her class with a motivational rhyme on "The Three Bears": "Papa Bear, Momma Bear, Baby Bear". She involved the entire class in singing the rhyme.

Soon after the motivation, she displayed a chart that depicted the entire story of "Curly Lock and the three bears". Then she asked the students to comprehend the story as per their understanding.

With the help of the chart, the teacher narrated the story to the students. She elicited the characters involved in the story, the plot, and the setting

The teacher read the lesson aloud with proper pronunciation and intonation. Children read/repeated the reading after the teacher. Using a few flashcards, the difficult words were explained with illustrations, she created a comparison activity in the Papa bear, Mama bear, and the Baby bear.

She divided the class into small groups and asked them to dramatize the story. Then she gave a worksheet to the students related to the lesson. The worksheet contained a jumbled story and tested the LSRW skills.

She gave an assignment to the class, A few hints were given for students to develop that into a meaningful story. They were also asked to write the elements of the story.

Case II: Class II B

Let us witness Mrs. Dhanalakshmi's class, who is affectionately known as Dhana Ma'am. She will be teaching the same lesson.

The students were divided into groups and the teacher showed them a picture of a forest via PowerPoint Presentation. The students were instructed to describe the picture.

The teacher asked the students to cut an outline picture of a bear as they were in their respective groups. Then she drew a few patterns on the board and asked the students to cut the bear into those patterns as shown. (z π v s). She then shuffled all the pieces gave them to the members in the same group and asked them to assemble them to form the correct bear image. It was exciting to see all the children working together.

The thought process was further provoked by posing them with a few questions –"How do you feel when you touch a block of ice? Warm water? Or a hot glass of tea"?

The children expressed their views.

The story was then dramatized in the form of a "Freeze Frame", where each group of children was instructed to play the role of a particular part of the story along with the settling.

Example: The children transformed themselves into things like a bed, a chair, a table & a bear after reading the part of the story assigned to them.

Then a chart was placed on the board and children were asked to write the names of their family members. Usage of "Mr.", "Mrs." and "Miss" before the names were explained and emphasized.

Recapitulation of the story was done by asking the students to choose a character from the story and one member to depict it through the drawing on the board while the other members instructed them with verbal or physical instructions.

The class was concluded by placing a few picture cards on the board. The teacher gave them a few word strips and asked the children to place the strips adjacent to the pictures.

The children were finally taken on a field trip to a forest the next day.

Conclusion

In Experimental learning cognition and competence and utilize in a meaningful manner are placed. It facilitates real-time coaching and learning for the students and accelerates the joy of learning from a novice to an expert, Art integrated learning, Activity Learning, Fun, games and studies, Sports Integrated Learning, Inquiry-based learning, Collaborative learning, Assessment as learning, Learning by doing and Best Teachers' method. Experimental learning promotes reflective thinking helps face new opportunities and challenges, and enables students to apply theoretical knowledge without any rigidity.

References

UNESCO. (2010). *Seoul Agenda: Goals for the Development of Arts Education.* http://www.unesco.org/new/fileadmin/MULTIMEDIA/HQ/CLT/CLT/pdf/Seoul_Agenda_EN.pdf

NCERT (2007). *National Curriculum Framework 2005*, Working Papers id: 1138, eSocial Sciences.

Sweeney, L.B. (February 22, 2016). *Learning to Connect the Dots: Developing Children's Systems Literacy.* URL: https://thesolutionsjournal.com/learning-to-connect-the-dots-developing-childrens-systems-literacy/

McGuire, S (2023). *20 Technology Tools to Engage Students in the Classroom.* URL: https://www.teachthought.com/technology/technology-tools-to-engage-students/

Chapter 10

Multidisciplinary Learning

S. Arulsamy and A. Benjamin

Introduction

> "By education, I mean an all-round drawing of the best in the child and man – body, mind, and spirit."
>
> – *Mahatma Gandhi (1937)*

Multidisciplinary learning is a learning methodology in which the curriculum of various disciplines is integrated to improve the learning experiences of the student community. As an example – a student of Engineering can opt for a subject in Humanities, related to social and civic services. The traditional method of a hierarchical educational structure is generally considered a one-size-fits-all strategy. It is walled by many aspects like repetition and memorization of facts that students care less about and retain at a reduced rate after assessments.

In a nutshell, the vision of the overall holistic development of a child is not nurtured, but compromised. This emerging competitive world requires prototypical educational logistics to promote a multidisciplinary approach to enable students to follow their interests and passions.

Multidisciplinary Learning in National Education Policy 2020

> "Mind is never a problem, but the mindset is"
>
> *(Narendra Modi, n.d.)*

 DOI: 10.1201/9781003504894-10

"Education must focus on "How to think rather than 'What to Think'."

(Godhwani, 2021)

Education is the backbone of the development of any nation, for a large country like India with diverse cultures and languages and huge regional inequalities, it is essential to have a unified robust national policy on education to achieve certain essential national-level educational goals.

The two-fold goals of education are

1. To enhance and empower employability and entrepreneurial youth skills (including creativity and innovation) to enable them to earn a living and contribute to national development.
2. To develop appropriate social and life skills (including values and ethics) that would enable them to lead a harmonious, happy, and ethical life in society.

Education has supreme importance in human life as it has the power to make the world a better place. National Education Policy 2020 is a comprehensive document and its main purpose is to improve the education sector by making it inclusive, holistic, multidisciplinary, and more fruitful. Undoubtedly, multidisciplinary education is extremely significant in this post-modern society of the 21st century. There has been a lot of emphasis on holistic and multidisciplinary education in the new NEP 2020.

Multidisciplinary and holistic learning is an effective approach that aims to fully stimulate all aspects of learning methods to cross boundaries between disciplines. In this type of learning the same topic, theme, or issue is studied from the crew point of more than discipline. Through access to infrastructure, trained teachers, and other newly introduced facilities through MERUs (Multidisciplinary Education and Research Universities), an all-inclusive teaching-learning that cuts across various subject areas and their relationship to the real world is established (NEP, 2020).

Advantages of Multidisciplinary Learning

1. **Comprehensive Learning Approach:** Education, is a complex world with many styles, values, and philosophies, multidisciplinary learning integrates the major elements of every discipline into a comprehensive pattern. It opens the pathway to different creative ideas and ways of understanding various branches of learning.
2. **Cohere with The Real World**: Multidisciplinary learning is a tangible way of understanding the world. Working collaboratively through effective communication across various spectrums is integral to multidisciplinary learning. The main focus of multidisciplinary learning is to create a better world by

identifying solutions through various subject areas by connecting with the real world.

3. **Synthesis of Ideas**: By emulating a multidisciplinary approach, students consolidate learning through a combination of ideas from many perspectives and choose the best alternative way to acquire knowledge. It also helps to build diversity of understanding concepts to positively impact learning.

The descriptors for transmuting the lives of students through multidisciplinary learning:

a. The educand explores multiple disciplines through a set of natural circumstances in which he can gain knowledge by himself
b. Multidisciplinary learning creates an attitude to the inquiry as most of the subjects are attached to nature-based studies
c. As the subjects are correlated, a student can immerse himself in learning
d. Thematic learning is used in the interdisciplinary curriculum
e. Educators act as coaches and mentors to aid learning in students

All educational institutions must focus on strengthening the skills needed for 21st-century successful adult life. Students must be empowered to think precisely and accurately to overcome the challenges in an information-driver world ruled by technology and enhance their innate potential.

Objectives of Multidisciplinary Learning

- To inculcate a rounded and holistic thinking pattern in all spheres
- To adopt an out-of-the-box approach to every problem and go the extra mile to resolve them
- To prepare Gen Z for the "World of Work" -with an arsenal of skills that will enable them to work in different environments
- To connect academics of different disciplines to make the learning process effective and efficient
- To synchronize and share the intellectual assets and skills that promote education

Multidisciplinary Strategies in Education

1. **Interrelate subject streams to Conceptualize:** Students must always be equipped to combine their knowledge and skills to help them transition from school to the workplace and the community. Combining the curricular objectives and different approaches to central themes from more than one discipline enables engaged learning in the teaching-learning process.

2. **Self-Driven Learning:** Inter-disciplinary learning gives students greater control of their learning and motivates them to acquire an array of skills that are needed for the holistic growth and development of the students. Through self-driven learning, their thought processes are stimulated and students become self-aware about their learning outcomes.
3. **Career-oriented Level:** Multidisciplinary learning being goal-oriented, helps students to co-ordinate different aspects of a problem and to reverse action or operation. This continued learning will prepare and shape them to excel in their careers in the future. This will also enhance their learning efficiency and broaden their thinking horizons.
4. **Cross-curricular Connections:** Multidisciplinary learning promotes the integration of subjects to highlight the interrelationships among different subjects. Various tools of learning are used to learn different subject areas to make cross-curricular connections. Students are therefore aware that their learning has an immediate real-world application.
5. **Higher Order Thinking Skills:** Based on learning taxonomies, higher order thinking skills (HOTS) is a type of learning that requires different learning and teaching methods. Through interdisciplinary learning, a student will be able to learn complex judgmental skills like critical thinking and problem-solving.
6. **Mixed Method Learning**: With the diverse approaches in the teaching-learning process, multidisciplinary learning acts as an orienting lens to engage students in a mixed-method study. It helps to diagram the process of learning through varied subject streams. Integrating various subject matters into a comprehensive and convergent design is one of the best-emerging ways to equip students with new knowledge and new skills.

According to idealists, there is unity in diversity in all objects in the world. Knowledge must be imparted to students when they feel that there is no one method of learning, rather whatever method of teaching aids in the exploration of knowledge is the most feasible one.

1. **Correlation:** A multidisciplinary teaching methodology widens the minds of students as they learn one subject from multiple dimensions. This knowledge enables them to understand the correlation that exists between different subjects. Their understanding and thinking patterns also undergo a change which equips them for better learning outcomes
2. **Interlink subject theorems:** A theorem is an obtained result that was already known to be authentic. In a multidisciplinary learning curriculum, subject theorems are one of the instruments or tools used by the educator to enrich the education to help achieve the aim of interdisciplinary education. Through interdisciplinary learning, the student is shown different elements of learning to guide the process of teaching and learning in a seamless manner

3. **Statistical Applications:** The history of education very clearly explains that teaching and learning methods must be developed and modified to meet the demands of the present generation. In multidisciplinary learning, the use of statistical applications in all subjects helps the students to arrive at reliable conclusions. This study enables them to learn through the derivation of valid conclusions.
4. **Habits of mind:** A multidisciplinary learning strategy helps students to inculcate the habits of mind for reading, writing, and thinking across various disciplines. In this context, students make connections with real-life situations, and transfer of knowledge through different contexts and perspectives are put together to co-learn the subjects.

Multidisciplinary Learning through Integrating Arts in Education

Integration of arts with other subjects means a blend of visual arts, performing arts, and literary arts that are integrated into the teaching–learning process. Arts, at the center of the curriculum help to clarify concepts and can provide a means to bridge the content of various subjects in a learner-centric manner. A joyful and creative learning environment is created through art-based learning. Art education encourages sensory explanation as it works with ideas and materials to create passion and expression which cannot be expressed in words alone. The abstract concepts are corrected through various forms, be it in the form of a song, a drawing a painting, or a performance. It enables students to develop multiple skills and abilities.

Through art integration in multidisciplinary learning, the three domains of cognitive, psycho-motor, and affective domains are fully utilized that help fulfill the need for competency-based learning and competency-based learning outcomes.

In multidisciplinary learning arts also address the diverse learning needs where every student can explore and experience a topic more deeply to express their innermost feelings through works of art. Art-integrated learning thus ensures that barriers are broken and students engage with each other for a better learning experience.

Table 10.1 gives a bird's eye view of an art-integrated checklist to analyze and understand if the goals of the multidisciplinary approach in arts integration are achieved. By melding the process together, a harmonious and integrated approach to learning can be accomplished.

Art Integrated Pedagogy

Education is directive in nature. The principal of the school must plan out its annual pedagogical plan of art education in detail at various levels and check how

Table 10.1 Arts Integration Checklist

APPROACH TO TEACHING		
Are learning principles of constructive (activity-built, experiential, evolving, collaborative, problem – solving and reflective) evident in my lesson?	Yes	No
UNDERSTANDING		
Are the students engaged in constructing and demonstrating, understandings as opposed to just memorizing and reciting knowledge?	Yes	No
ART FORM		
Are the students constructing and demonstrating their understandings through art form?	Yes	No
CREATIVE PROCESS		
1. Are the students engaged in a process of creating something original as opposed to copying or parroting?	Yes	No
2. Will the student receive their products?	Yes	No
CONNECTS		
1. Does the art form connect to another part of the curriculum or a concern/need?	Yes	No
2. Is the connection mutually reinforcing?	Yes	No
EVOLVING OBJECTIVES		
1. Are there objectives in both the art form and another part of the curriculum or a concern/need?	Yes	No
2. Have the objectives evolved since the last time the students engaged with this subject matter?	Yes	No

art can be integrated with classroom learning. The essence of art education must be to reinforce art as a subject and as a tool for learning.

For implementing this in classrooms, the teacher may think and prepare their plans while keeping in mind the objectives and the learning outcome. Art experiences in the form of ice-breaking, follow-up exercises, and linking of art experiences with the concepts or themes ensure active learning where both the content and art are integrated well to give seamless learning in both areas.

Examples of Multidisciplinary Learning Approach

Primary Students

Teachers understand the importance of learning by doing a few activities to clarify the distinctive nature of multidisciplinary learning for the primary section

Activity 1 – Multidisciplinary Learning

Primary Section	Mathematics
Class	IV
Topic	Charts and Wheels

Game Plan – 1

The objective is for students to differentiate between the shape of a circle and an oval form. The teacher blindfolds the students and asks them to identify the circular-shaped objects from a group of other objects belonging to different shapes. Probing questions are asked to analyze the difference between a circle and an oval shape. Students then conclude by learning that a circle is made up of curved lines.

Game Plan – 2

The template of a circle is given to each student according to their height order who are then divided into four groups. Each group is assigned to make a creative object using the circular template. Objects like buttons, chapatti, wheels, and bindi are made by students to help achieve the learning outcome.

Game Plan – 3

The activity was conducted to learn the center, radius, and diameter of a circle. Students were given a rectangular-shaped newspaper with a set of instructions to make it into a perfect circle. The first step is to fold the sheet in half and then in half again the other way. A triangle is obtained by folding it again in half. The shape of a cone is achieved by doing another fold to the paper. The final step is to cut off the top in a curved shape and the perfect circle is ready. Now the students learn to identify the center, radius, diameter, and circumference. In groups of four, they mark the center, diameter, radius, and circumference of their circle in different colors. Students enjoy this fun-filled activity and the learning outcome is also achieved.

Upper Primary Students

Activity 2 – Multidisciplinary Learning

Upper Primary Section	Science
Class	VI
Theme	Weaving Wonders
Topic	Fiber to Fabric
Sub-Topic	Production of Cotton
Form of Art Activity	Weave a Paper Mat

Game Plan – 1

Converting a raw material of fiber into a finished product of fabric is demonstrated with the help of the activity of a paper mat. The teacher first lectures about the cultivation of cotton and its requirements for its growth. The difference between natural and synthetic fiber is then explained to the students to make them understand the types of fibers. To make learning more creative, a paper mat is made and decorated with fabric painting where cotton cloth fabric paints and other materials are used to embellish the mat.

Another activity involves students sharing their thoughts about Mahatma Gandhi and his ideologies about wearing Indian-made clothes made of spun yarn using the charkha and was a part of the Independence Movement.

Inter-disciplinary Approach for Varied Subjects

In *English*, a write-up about "Coimbatore – the Manchester of Tamil Nadu" – popular in cotton production highlights the importance of cotton clothes worn during the summer season. A role-play depicting the difficulties faced by the weavers also is discussed as to why weavers give up their traditional work. Through Social *Science*, a Group discussion about the ancient type of clothing used by our ancestors like the bark of trees leaves of trees animals' skins is discussed and informed to the students. In *Mathematics*, students are encouraged to draw a charkha and point out the center, radius, diameter, and circumference of the charkha which helps them to get a detailed understanding of the topic Circles.

Extended Learning

With the help of an animated video, students are taught how cotton from the field is taken to the factory and then made into fabric. Pie charts are then made by surveying the number of textile industries in India, where students learn a concept in Mathematics too. Thus, several academic disciplines are combined to teach a topic to the students.

High School Students

Activity 3 – Multidisciplinary Learning

High School Section	Science
Class	IX
Topic	Structure of the Atom
Sub-Topic	Sub-Atomic Particles
Form of Art Activity	Model of an Atom

Concept: Sub-atomic Particles

The concept of the structure of an atom and its subatomic particles such as protons, neutrons, and electrons are taught on a paper plate periodic table. Construction paper or color papers are used to represent "P" (Proton), "E" (Electron), and "N" (Neutron). Coins and dice can also be used to represent them. Students enjoy this type of learning as complex concepts are learned quickly with the help of an activity.

Inter-disciplinary Approach for Varied Subjects

The concept is discussed through the subject of *English* where Role play is conducted through an open discussion. Students play the role of eminent chemists like John Dalton, JJ Thompson, Ernest Rutherford, and Niels Bohr. They discuss the model of an atom where students imitate the character and behavior of these personalities. Essay- writing is also an activity where ancient and modern philosophies of atoms are written. In *Mathematics*, using a bar graph, students learn how to show the valency of the first ten elements. A map study in *Social Science* is done to learn the places where radioactive elements like uranium and thorium are found. The places are then marked on the map to learn the nuclear power stations. This map study enables students to understand the concept of atomic theory. Through Art, a schematic atomic structure model of carbon, oxygen, sodium, and chlorine is made with the help of coin beads or clay balls of the same size.

The teacher narrates a story to the students. The story requires the students to build up a miniature model with mathematical specifications.

Activity 4 – Multidisciplinary Learning

Secondary Section	Mathematics
Class	IX
Topic	Surface Areas and Volumes
Form of Art Activity	Modeling

Narration

> There was a boy named Ashish who lived in an apartment. There was a park which was inside the gated community. The park was circular. Children used to play in the park every evening. There was a cuboidal fish tank in the park which had colorful species of fish. But now due to the pandemic, children are not allowed to play in the park. Little Ashish missed his playtime. One fine morning when he was sitting aside his window, he got the idea of making a miniature of his house and park and inviting his friends to play through an online meet.

The teacher asks the students the shapes required to make the miniature of the house and park. The teacher also discusses the materials required to make the miniature.

This methodology is taught in Maths with the use of three-dimensional shapes like a cone, sphere, prism, cylinder, and cube are used to build a miniature house. The target learning outcome about the surface area and volumes are learned. It is a fun way to integrate math and process art. Students are made to sit in groups and discuss the shapes needed to make the model of a circular fish tank. The surface area, the quantity of water, and the measurements needed are reviewed and answers are explored for the same.

Inter-disciplinary Approach for Varied Subjects

English teacher teaches the lesson "The Fun They Had" through online and offline modes. Few students attend the classes through both forums. A differentiation and a discussion take place to highlight the pros and cons of classroom learning V/S online learning. In *Physics*, the concept of circular motion is integrated with the circumference concept by finding out the motion of the body along the closed path. For example, in a park, there comes an old man with a dog, find the distance covered in the circular walk path of diameter 100m. In chemistry, students can be asked to find the volume of the concentration in a solution which is related to the concept of volume. The concept of an online classroom or online meeting with friends is integrated with *Information Technology*. By drawing the two-dimensional form of the miniature home, Art is integrated into learning. The concept of finding volume is integrated into the lesson தண்ணீர் and students are guided to apply the concept of finding the volume of water in a given container through the subject *Tamil*. Multidisciplinary learning achieves its objectives by synchronizing the intellectual assets and skills needed for learning. Therefore, *Holistic Education* is an integrated learning methodology that adopts a comprehensive approach to teaching where the emotional, social, and psychological needs of the students are fulfilled.

Conclusion

The multidisciplinary approach described in this study provides a greater value to the students where the curriculum highlights real life that is multifaceted than being compartmentalized into a neat package. Real-life problems are complex and so, no single discipline can resolve the issues. It is now explored that interdisciplinary forms of learning are widely accepted and growing in abundance and stature throughout higher education.

Educators engage students through varied learning techniques and help them to increase their knowledge and skills through multidisciplinary methods of learning.

Multidisciplinary teaching increases student learning. Engaging students and helping them to develop knowledge, insights, problem-solving skills, self-confidence, self-efficacy, and a passion for learning are common goals that educators bring to the classroom and interdisciplinary instruction and exploration promote the realization of these objectives.

The concept of creativity has expanded and changed in recent years. Using materials that mimic real-life situations and multidisciplinary collaborative projects is becoming a mandatory part of curricula in all fields of education. Experiences that are real provide a structure and shape for many scholars who have attempted to understand learning with a multidisciplinary outlook.

Multidisciplinary learning prepares Gen Z for the "World of Work" -with an arsenal of skills that will enable them to work in different environments, connect academics of different disciplines to make the learning process effective and efficient, and synchronizes and share the intellectual assets and skills that promote education.

References

Gandhi, M. K. (1937). Literacy in itself is no education. Harijan, 31-7-1937

Narendra Modi (n.d.). URL: https://quotefancy.com/quote/1619567/Narendra-Modi-Mind-is-never-a-problem-Mindset-is

Godhwani, R (December 01, 2021). Make them lifelong learners. The Hindu. URL: https://www.thehindu.com/education/teach-students-how-to-think-rather-than-what-to-think/article37782981.ece

National Education Policy (2020). URL: https://www.academics4nation.org/

Chapter 11

Empowering Team Collaboration Using Digital Technologies

An Innovative Teaching Styles in Higher Education

Antara Dey, Nil Ratan Roy, and Raj Kumar Pal

Introduction

A teacher constructs the learning environment for the students suitable to their learning needs as per the demand of the situation. The suitability of a teacher's teaching style is the most determining factor of students' readiness and motivation to learn (Shaari et al., 2014). Teaching style is general pattern of teaching behaviour created by particular set of strategies used by a teacher. (Bhardwaj, 2009). The teaching style is one of the prime factors that influence the learning environment and the success of the teaching-learning process (Dincol et al., 2011). The classroom is a social laboratory wherein the teacher must build a web of communication that becomes a medium for bilateral transactions between the teachers and students. The role of a teacher in the digital era is very important in higher education. In the 21st century, teaching joyfully using digital technologies is a challenging task for a teacher.

Due to individual differences, teachers have different styles of teaching in higher education. Naturally, teaching style may be the outcome of the different personality characteristics of the teachers. Teaching is the profession that creates all other

DOI: 10.1201/9781003504894-11

professions. To attain good results in teaching, it is necessary to appoint good teachers. An effective teacher is competent, resourceful, can manage the classroom, has a good interpersonal relationship with students, colleagues, and the head of the institution, possesses good knowledge of the subject matter, and is disciplined, mentally, and emotionally strong. They can inculcate the same to learners. Teachers need to be able to work in teams and cooperate with colleagues and parents of the students (Saranyadevi, 2018) in higher education. There are several characteristics like teaching behavior, motive for teaching, favored teaching styles, content knowledge, and capability to classroom manage; their characteristics affect teacher performances in the classroom and the teaching-learning environment as well (Khandaghi and Rajaei, 2011). Teachers use different teaching styles to attempt to create classroom social-psychological climates (Kumari, 2008). By adopting a certain style, the teacher models a certain role for students for effective student-centric teaching in the digital era.

The role of teachers in the 21st century is that of facilitator, open-minded, co-operator, mediator, and collaborator, and they need to be techno-pedagogically skilled at technological momentum. The Flipped classroom, social media, and smart classroom provide a collaborative platform to grow higher education students as per the need of the hour. As per the National Education Policy (2020), the National Educational Technology Forum provides a platform to exchange ideas, and the use of technology provides an assessment, planning, and administration platform in higher education. There is a huge paradigm shift from simple chalk-and-talk autocratic teaching to student-centered, constructivist digitalizing approaches. Teachers should provide knowledge to develop students' creativity, curiosity, and motivation. Team collaboration in today's digital era needs to use wikis, weblogs, and social networks to work collaboratively and share information to reach a common goal. Higher education institutions are facing opportunities and challenges for digital revolutions. However, the teacher's role should not be replaced by digital technologies.

Concept of Team Collaboration

Team collaboration means when people work together with complementary skills to achieve a common goal and for which they should be mutually accountable to each other. Effective team collaboration is very important for any success. It is the action to work together to get the maximum outcome. Team collaboration is an interaction between different cultures and d technology such as audio and video conferencing emails and social media. Team collaboration in higher education where with the support of teachers students work together to create and explore innovative ideas to fulfill a common purpose. It encourages the active engagement of students in active knowledge constructions. Teachers bridge a link between knowledge and community. Team collaboration helps in conversational learning (Pask, 1975), academic knowledge(Laurillard, 2001), condition for deep learning (Marton and Saljo, 1997; Entwistle, 2000), and knowledge construction(Scardamalia and Bereiter, 2006). Teachers mainly focused on communication using social media (Hiltz and Turoff, 1978) through discussion among teachers and students.

Characteristics of Team Collaboration

There are some characteristics of team collaboration in higher education. These are discussed below

i. **Integration:** It means students learn together by integrating their knowledge within a small group to achieve a common goal. A teacher should accept the integration and, at the same time, students need to accept this positively.
ii. **Communication**: Communication is very crucial for team collaboration among students, as it helps to build a sense of interrelationship between students. Humans are agents of society. Individuals have different intellectual abilities so if all people work together they can adequately share their ideas, and outcome will be better. Team collaboration help create in a sense of belonging through communication. Sharing insight and ideas helps to get better success.
iii. **Accountability**: Team collaboration helps in accountability in terms of performance. Team collaboration helps in an extremely cooperative environment and assures a consistent outcome. No one criticizes an individual's failure, rather celebrate success together.
iv. **Structural Execution**: teachers who believe in team teaching, first of all, prepare a roadmap on proceedings, future risks, and delays.
v. **Measurable goals**: Teacher facilitation on setting measurable, realistic goals. This keeps every individual's effort going toward the desired outcome. Goals should be common, transparent, and focused. A teacher should appreciate each individual's goals for fulfilling a common purpose.
vi. **Balanced motivations:** Teachers can motivate students, and generate values, in whatever they work. Risk-taking behaviors are encouraged. They are always looking to generate a drive for students' success.
vii. **Constructive conflict:** Teachers gave an open chance to share ideas and criticize students' activities. Then, it is quite natural that arguments and conflicts may emerge. So, here, the teacher's role is to understand the situation, and such constructive conflicts are necessary for students to fine-tune their ideas.
viii. **Diverse Team Members:** Every individual student has their capability, experiences, and knowledge. Diverse students in a class can put diverse ideas. Each student has their skill. A variety of students in an inclusive classroom can bring creativity and a broad range of ideas. It creates a strong interpersonal relationship with the teacher and other fellow students. This integrated collaborative teamwork helps to achieve more success.

ix. **An organized good leader:** A teacher plays the role as a leader that they can get trust and respect. This teacher can be teaching such that they can be able to hold the team together and should be able to set the pace, offering encouragement and motivation to work together collaboratively. The effort of a good leader is essential for every success.
x. **Joyful learning:** The teacher's role is also important for learning joyfully. Without any fun, can leads lack productivity? So, it is important to put a bit of enjoyment in a multicultural, inclusive class. Whether the teaching is online or offline, it depends on the teacher's teaching style to work with the help of different technologies.

Elements of Team Collaborations

There are certain major elements of team collaboration are discussed below

i. **Motivation:** Motivation means the drive to work hard. The role of a teacher is to create that drive. A teacher's teaching style should go in that way to motivate students extrinsically and intrinsically. If students are motivated to do work then the success of the group is possible.
ii. **Communication:** communication is another important element of team collaboration. Communication is possible without collaboration, but collaboration is not possible without communication. In team collaboration, a group of students can take responsibility for achieving success. In every step of team collaboration, strong communication is very important for team achievement.
iii. **Diversity &Engaged employees**: A classroom consists of several students from different socio-cultural backgrounds. Teachers need to teach properly in an inclusive classroom where different diverse students are present. The role of a teacher is to engage them in their work. In an inclusive classroom, diverse ideas help to achieve more success.
iv. **Sharing attitude:** Teachers' attitude toward sharing knowledge is very important. So that students can also follow the same rules with their friends. In team collaboration, it is very important to share knowledge and ideas with others to get more creative innovations. This type of environment helps to get more success.
v. **Support& trust:** The teaching styles of a teacher should be the way that helps to develop support and trust among the students. These are two essential components of team success. When everyone supports each other without finger-pointing, it helps to get more outcomes.
vi. **Conflict management:** Teachers should have the ability to solve the problem. There are several types of students present in any classroom. There may be some different opinions of each student and one student may finger-point at others' work. It may create some conflicting situations among the students. A teacher should address that type of conflict in a constructive way. So, conflict management is very important for team collaboration.

Purpose of the Study

The purpose of this study is the conceptual understanding of team collaboration in higher education with the main focus on teamwork in a collaborative way and a healthy team learning environment.

Objectives

The major objectives are to summarize the characteristics and elements of team collaboration, to find out the digital activities involved in learning through team collaboration, to explain the role of a teacher in using techno-pedagogic skills for effective team collaboration, to discuss different digital strategies to enhance team collaborative culture in Higher Education.

Methodology

The present study is mainly descriptive and analytical. This paper is mainly based on different secondary sources. These are taken from different books, articles, journals, and magazines.

Importance of Team Collaboration in the Digital Era

There is a phrase that "No man is an island." It means that no single person is self-sufficient. Always a group of individuals is more powerful than a single person. Teamwork helps employees to share knowledge and work effectively. Team collaboration using digital technologies is very important to enhance optimized workflow processes, Optimized outcome interpersonal relationships, and decision-making skills to fulfill a common goal. However, with the changing digital world, the role of teachers in team collaborative ways of teaching is also changing in the workplace. With this transformed working culture nature of the digital landscape and team collaboration also have become challenging.

i. **Leadership skill development:** Mainly, the teacher leads the rest of the team for more effective work. Team members help everyone to work together at every stage. This team collaboration helps the student to inculcate leadership skill development. Together, they can learn better.
ii. **Impact of Technology on Teamwork and Collaboration:** Technology and digitization can be said to be a game-changer. In the 21st century, every person is connected with digital technologies. Today's classrooms are also transforming into smart classrooms full of technologies. In the same way, teachers in the workplace are also techno-pedagogically skilled. Apart from physical team collaboration in Google Classroom teachers also need to be

techno-pedagogically skilled. Workers can do work from any comfortable place.

iii. **Improved Internal Communication and Collaboration:** Team collaboration is important for improving internal communication and collaboration so that work should be more effective. Apart from physical team collaboration and communication, different collaborative instant messaging digital tools at a virtual pace can improve efficiency and transparency in the workplace.

iv. **Improved Productivity:** A team can work with more than a single person within a stipulated time. Digital collaborative tools help save time and improve productivity. Teachers and Students can work from their comfortable places by using digital collaborative tools. It helps in getting content easily through using the internet. Through all these, it helps to improve productivity and work culture.

v. **Team Interaction:** In the digital era, team collaboration is possible through using different digital tools, which helps in good interpersonal relationships. Team collaboration helps in team communication, which leads to good interpersonal relationships. Different digital tools like Zoom, Microsoft Teams, and Google Meet help to work collaboratively from a comfortable place. They can interact, text, or even video to work collaboratively. It enhances good interpersonal relationships.

vi. **Knowledge Management:** Different knowledge management software helps in effective collaboration. It helps to store different internal documents as well as different external materials safely. These types of tools help everyone to work efficiently.

Digital Activities Involved in Learning through Team Collaboration

Teachers can play a very significant role in the teaching-learning process by using different techno-pedagogical skills. It means the competency of teachers in using pedagogy. Beaudin and Hadden (2004) said techno-pedagogical skills are very important skills for teaching-learning situations. It is based on constructivist pedagogy (Sondhiya, 2015). The teacher's role is to possess the techno-pedagogical skill to imbibe modern multimedia for the current teaching-learning process. Students use multiple sensory modalities through modern technology to learn the practical aspect of theoretical knowledge (Ali, 2018). This type of learning using digital technologies helps the students to get the latest information, inspires students for student interaction, assists in building practical knowledge, helps to develop creative ideas through using different technologies, organizes record and report information, and helps to enhance collaborative and cooperative learning environments. Teachers' team collaboration style helps in different teaching-learning processes. They help students' communication skills, learning at their own pace, real-world information, authenticity in learning tasks, pragmatic and advanced information, independent self-learning, commitment to learning activities, and help in higher-order thinking. In this era of digitalization, teachers of higher education institutions implement the latter for effective learning outcomes among learners. It is related to different teaching skills.

The teacher's role in technical education is very much crucial. The teacher's role is lifelong learner and innovator. Teachers can create an environment through virtual laboratories, e-learning resources, open educational resources, etc. in that students become lifelong learners. Teachers' roles should provide an advanced level of knowledge. Techno-savvy teachers must empower and enable themselves for technology improved teaching-learning process. Teacher and students should sharpen their knowledge through digital technology. Content knowledge is one of the most important aspects of digital technology. In the 21st century, there is a need to develop technical skills that are needed everywhere. Now, in the 21st century, the teacher's role has changed with the advancement of technology, and they are trying to use it in every section and it is transcendent into the four walls of the classroom. In higher education, students should promote self-learning. The teacher's role is facilitator and manager for all-round development in the digital environment. So, in that way, the learner utilizes optimum online resources. There are digital activities involved in learning through team collaboration are discussed below.

i. **Virtual Team Meetings:** Teachers should have virtual team meetings to develop team collaborative skills among students. It is essential for students of higher education because they can do more by meeting each other and sharing their knowledge at the same. This will provide a better result. It is a very essential activity for team collaboration.
ii. **Live Remote Office:** It is also another important activity for team collaboration. Virtually, students can meet through Google Meet or Google Classroom. The teacher starts the meeting virtually, and students work and learn collaboratively from a comfortable place. It seems they are working together with a physical presence. So, they can work online in video audio mode on can essentially join in a workspace. Here, all members work together and get motivated and focused on generating productive work.
iii. **Slack Extensions for Team Building:** Donut is one of the Slack extension virtual team-building processes. Donut pairs students to converse with every one to four weeks. So that, the bonding of every student remains the same. These meetings are mainly for creating social relationships with students apart from academics with the help of technology. Later, students feel more active in work.
iv. **Capture Important Events:** Sometimes, students are unable to **attend** all the classes physically, and it is also quite difficult to see other students' hard work. So, here, student does not feel left from their work and can attend classes

virtually. In the video, they can congratulate the students who perform best in terms of their hard work and get a reward. So that other students also get motivated from there to achieve more success.

v. **The Deserted Island Scenario:** It is a one-off virtual team-building game. Here, team members are given a desert scenario with seven objects, but they can choose one. Objects should be as strategic as possible so that they can choose three and think and plan accordingly. Then, split the team to choose objects through collaborative discussions. These will create higher education students working together and get better results.

Role of Teachers in Using Techno-pedagogic Skills for Effective Team Collaboration

Techno-pedagogical skills are one of the most important competencies of teachers to use technology successfully. Teachers can play a very significant role for students as per the changing needs of society with the help of effective techno-pedagogical skills (Gloria and Benjamin, 2014). Different techno-pedagogical activities like e-learning content, online Teaching–learning materials, and electronic dictionary techno-pedagogical competency promote higher education (Koushlesh and Yadav, 2015). It assists teachers in a techno-pedagogically skilled environment. Teachers' innovative skills and critical thinking abilities depend upon techno-pedagogical competency. Teacher's innovative skills and techno-pedagogical competency will help students' development (Yildiz, 2017). There are some mediums like Zoom and Google Meet through which the online teaching-learning process is conducted. At present, a transformation has happened in the education system from traditional teaching to technology-oriented teaching. Here education becomes topsy-turvy. Mainly, the chalk-and-talk method of knowledge and teaching process changed to a digital platform. Digital technologies are the main pathway of education. Therefore, the teaching process remains to continue in the digital world. The online teaching-learning process is a blessing for teachers as it allows them to interact with others. Teachers need to be competent in their pedagogy. Human resource development centers under UGC create several courses for higher education. New-generation higher education students are mainly comfortable with technology-oriented teaching rather than the one-way dialogue method of teaching by teachers. Today's education system is interactive and student-centric so that students can participate in the whole process (Ahmed, 2020). Blended types of teaching are more important, whereas both type online and offline methods of teaching are given importance. The different techno-pedagogical skills of teachers through the use of different technologies are very important nowadays in higher education. Self-learning by students of higher education has been given more focus (Vajargah, Jahani, and Azadmanesh, 2010).

The teacher should be techno-pedagogically skilled like developing different linguistic abilities and improving the teaching-learning process. The major significance can mention different aspects relating to techno-pedagogy, which are directly and indirectly helpful for higher education students in terms of encouraging auto-learning ability, increasing enrollment, improving the examination process, supporting research activities, stimulating cognitive learning, improving life skills, building aesthetic sensibility (Thakur, 2015). Different professional development courses and study materials are used through techno-pedagogical skills. It also helps to develop life skills among students (Vijaya, 2017). It helps to use knowledge of the competency in a meaningful way (Ali, 2018). The shift from face-to-face learning to virtual classes is one of the most important solutions for adopting techno-pedagogical skills. There are certain challenges like distance, scale, and personalized teaching-learning process for virtual teaching (Dhawan, 2020). In different higher education institutions, teachers use different online tools such as Google Meet, Zoom, Edtech, and Webex (Basilaia and Kvavadze, 2020) for team collaboration. Teachers with high techno-pedagogical competency for auto-learning for the students through learning courses like MOODLE and MOOC (Chattopadhyay, 2020). Working collaboratively in a team is very important to learn more and to get success. It can be enhanced with the help of technology. So, technology-based different pedagogical approaches unconditionally fulfill the requirement of the whole situation. The importance of this type of skill helps to get more success in the field of education. The teacher's techno-pedagogical skill needs to be enhanced.

Digital Tools Used for Team Collaboration among Learners

The developing technology can impact any institution. Virtual collaboration allows everyone, irrespective of their geographical location. In higher education, students can learn at a comfortable pace. They feel that they seem present in a group. Teachers of higher education should play the role of facilitators. The teacher's teaching style should be one in which team collaboration is enhanced through the use different tools. Several tools of team collaboration in higher education are discussed below.

i. **Microsoft Teams:** It is one of the most important tools for team collaboration. It is a virtual team collaboration tool that helps in video conferencing and other activities among the students. It helps to store different files and plenty of application integration with other required workplace programs. It helps the students to join in any classes or meetings for academics. It consists of different programs like PowerPoint, Word, Excel, etc.

ii. **Zoom:** It is another tool for team collaboration. Zoom is software where audio-visual conferencing is possible. This product offers a single communication suite for different meetings, chats, and productive collaboration in the workplace. It helps to create virtual interactive sessions among higher education students. The product mainly works on Mac, Linux, iOS, and Android. This platform helps anyone to attend meetings through desktop and mobile devices from any comfortable place.

iii. **WebEx:** It is also a virtual team collaboration tool mainly under Cisco Systems. It is also one of the most popular tools for video conferencing, online meetings, webinars, seminars, and workshops for higher education students. Students pursuing higher education may utilize this highly adaptable and user-friendly platform according to their needs.

iv. **Proof Hub:** It is a perfect project management and online collaboration tool. It helps students to do teamwork and stay task on top of their side. Anyone can break projects into different tasks and smaller subtasks to assign them directly to team members. Proof Hub helps it in an easy way to take it in streamlining projects and schedules.

v. **File stage:** one of the most important online collaboration tools is the file stage. It helps to manage different types of content review processes, easy feedback provider on the video image and PDF, etc. It also helps in the creative management of the team. It saves time.

vi. **Fresh desk:** It is a robust tool for team collaboration. It helps in extraordinary customer support. It helps to do teamwork effectively. It is an online tool through which it is possible to invite team members and work together collaboratively.

vii. **Intercom:** It is one of the interactive collaborative software which helps the team in a very interactive way. It helps in real-time chat to share and integrate their knowledge to support students' efforts.

viii. **Troop Messenger:** It is a powerful, flawless, team communication and collaboration tool for audio and video calling, along with voice messaging and chat messaging for students from a comfortable place. It is available on different Browsers and Androids. It is self-explanatory and helps in screen sharing and file sharing for the students.

ix. **Chanty:** It is an online collaboration platform for team collaboration and improves the team members' healthy relationships to collaborate team members. It will help make team communication transparent and accessible. It is a kind of team that helps in high-quality audio-video calls for higher education students.

x. **Google Docs:** It is one of the most important tools for team collaboration with a small and big team. It helps to share, edit, and publish documents in all kinds of spreadsheets and different types of presentations. Formatting of documents with Google Fonts typefaces is also possible here.

xi. **Slack:** It is one of the most foundational team collaboration tools. It helps to communicate and collaborate in different teams easily. It helps you to set one standard for online team communication.

xii. **Share File:** It is a good collaborative tool for storing and sharing different types of files that are mainly prepared for business purposes. It helps different students with professional development.

xiii. **Flock:** It is a very easy way to use virtual collaborative tools. Students can work together and interact with themselves. As a facilitator, the teacher uses this app to start the meeting, incorporates student ideas to spark new ones, and supports teamwork.

xiv. **Join. Me:** It is a platform for hosting a free meeting or joining a meeting to enhance different types of online meetings for higher education students. It is very important for team collaboration.

xv. **Git Hub:** It is a very good process of documentation and helps us to work together to solve different types of problems. It helps them learn the problem from each other along the way. It helps to manage well-maintained documents along with writing codes.

xvi. **Bit.ai:** It is one of the most important smart tools for team collaboration platforms for higher education students to create interactive insights. In this documentation software, team members can invite their other team members depending on the collaborative work.

xvii. **Click Meeting:** It is a perfect tool for effective collaboration and team communication. It is web conferencing to work collaboratively. Here, students participate effectively. It helps to easily engage your audience and allow every individual student to participate here in an effective manner.

Benefits of Team Collaboration

The use of technology for teaching is very important in the 21st century. The main aim of online collaborative learning is not to replace teachers but to use technology to help teachers and learners do work more effectively. The teacher's role should be to enhance team collaboration. These are some benefits of team collaboration:

i. **Problem-solving skill:** Teachers' teaching styles are in such a way that they should enhance team collaboration. In an inclusive classroom, so many students come from different socio-cultural backgrounds. Each student has individual differences in terms of their thinking, ideas, creativity, and intelligence. Online or offline team collaboration among students helps them solve problems effectively and efficiently through sharing their ideas.

ii. **Spurs innovation:** The diverse expertise of the teacher who plays the role of the facilitator will foster more innovation. Different students share their ideas for continuously updating with new information. It helps in generating new creative ideas.

iii. **Boosts skills-sharing:** It is a tremendous opportunity for a teacher to work together with students collaboratively to get better results. If the students can share their innovative ideas then it will create a constructive environment. So, team collaboration helps to enhance skill-sharing abilities. Team members work in a diverse environment and have focused decision capability, leading the team to a swift achievement of a common goal.

iv. **Adaptability:** If the students are working in a group then they can adjust to every situation with them, rather they do not feel lonely. Working in a team is far better than working individually. Like problem-solving is the natural result of an adapted collaborative team.

v. **Open communication and participation:** Team collaboration gives a chance for every member of a team to communicate freely and help to Participate in a group. Communication is an essential component of team collaboration. Without collaboration, communication is not possible. Team members work together to achieve more success. The level of participation helps in innovations.

vi. **Goal alignment:** In team collaboration, a common goal emerges from here through a combination of individual and team-driven efforts. It helps students to combine their abilities and knowledge to achieve a common goal. It helps to achieve more goals wide support in a team, and increased productivity.

vii. **Engagement:** If the students are working together then they are much more focused on their work. It supports students who can share ideas and align for team goals. If the students are engaged, then they will enhance productivity, retention, happiness, and success.

Team collaboration helps to develop more intellectual skills such as critical thinking, analytical thinking, and evaluation, which are the major needs in this digital age. Try to be flexible. In team collaboration, there are many minds are mixed to create innovative ideas and success.

Challenges of Digital Technologies for Team Collaboration

There are some challenges to using digital technologies for team collaboration. In the 21st century, teachers are not well trained for team collaboration using digital technologies. The teachers who are habituated to traditional teaching are not techno-pedagogically skilled. These are some challenges of team collaboration:

a. **Information overload:** In team collaboration through digital technologies, there is a need to digest and adopt digital technologies, which is becoming a challenge. The primary cause of this is a lack of understanding about what constitutes productive teamwork. Everyone strives to impart successful teamwork in their own unique manner

b. **No immediate results:** To make an effective team, collaboration is a long-term process. It is no matter how to use different tools for noticeable results. Getting no immediate results is not possible from team collaboration. Students need to give more time to get results of effective team collaboration.
c. **Collateral costs:** Teachers need to train properly for collaborative ways of teaching. Mainly, collaboration takes effort and money. There are so many knowledge gaps between the team or even within the team. Some people also do not know the process of using different digital technologies properly. There are so many places where there is no electricity. It is a very cost-consuming process.
d. **Lack of motivation:** There is a lack of motivation sometimes for team collaboration. There are some people with negative attitudes and pessimistic views. This type of teacher cannot motivate students to work collaboratively. It is extremely important to identify those people otherwise; the whole society will be hampered shortly. It decreases the whole team's productivity and undermines the morale of the team. It does not scale easily and requires a highly knowledgeable and skilled instructor.

Techno-pedagogical skills help in hybrid learning rather than simple education, especially in higher education. But, there are several challenges like poor infrastructure, poor awareness lack of trained teachers in practicing techno-pedagogical skills, shortfall of coordination, deficiency of research, and developmental aspects. Team collaboration is very helpful for more success but at the same time, but it is also very challenging.

Conclusion

The education system is changing with time. This cannot be identified exactly, but it is a continuous process where everything changes naturally. In recent times, ancient traditional teacher-centric education has changed into a student-centric system where interaction gets more important. Pedagogy means the art and science of teaching, whereas teachers need to be techno-pedagogically skilled in this digital era. The teachers are also inclined toward the techno-pedagogy, which has also increased. Techno-pedagogical exposure helps teachers to reflect critically and have a never-ending developmental journey. A teacher should need to be well aware of different digital technologies for the virtual world. Teachers need to develop different soft skills also. Nowadays teachers and textbooks are replaced by different digital resources. New-generation students are habituated to digital technologies and smart classrooms (Shah, 2014). Traditional classrooms have changed into smart classrooms full of digital technologies. So, we need to reevaluate the teachers and train them accordingly. The nature of students and the nature of the classroom also changed with time. So, we should construct our education system in that way. The teacher's role is like friend and facilitator. Teaching through different groups is more effective than teaching in a single individual. Team collaboration is one of the most important approaches to teaching in today's generation. A teacher should accept and teach in such a way that enhances team collaboration. Teachers need to help students think, reflect, and implement their knowledge to enhance team collaboration. Although there are so many challenges to team collaboration, it is possible to conduct more successful collaborative learning is possible in both face-to-face and online learning contexts. However, there is so much evidence that shows that collaborative student-centered interactive learning is more important to meet the needs of the digital age. Mainly at the higher education level students are more mature and interested in team collaboration to learn in a group. So, in this digital era, there is a need to enhance team collaboration for higher education.

Bibliography

Ahmed, A. (2020). Synchronizing pedagogy and technology in post COVID scenario. *International Journal of Creative Research Thoughts*, 8(6), 243–246.

Ali, M. (2018). Techno-pedagogical competency of teachers: An area of concern in the 21st century higher education. *Review of Research, Luxmi Book Publishing*, 7(11), 1–5.

Basilaia, G. & Kvavadze, D. (2020).Transition to online education in schools during a SARS-CoV-2 coronavirus (COVID-19) pandemic in Georgia. *Pedagogical Research*, 5(4). www.ijcrt.org © 2020 IJCRT | 8(11) November 2020 | ISSN: 2320-2882 IJCRT2011433 International Journal of Creative Research Thoughts (IJCRT) www.ijcrt.org 3669

Beaudin, L., & Hadden, C. (2004). Developing technopedagogical skills in pre-service teachers. In *E-Learn: World Conference on e-learning in Corporate, Government, Healthcare, and Higher Education* (pp. 492–498). San Diego, CA: Association for the Advancement of Computing in Education (AACE).

Bhardwaj, M. (2009). A study of teacher effectiveness about teaching styles and personality types of secondary school teachers (Doctoral Dissertation, Panjab University). Retrieved from http://hdl.handle.net/10603/84841

Chattopadhyay, A. (2020). Post-Covid language teaching-teacher as a techno-pedagogue. *Journal of Emerging Technologies and Innovative Research*, 7(8), 415–424.

Dhawan, S. (2020). Online learning: A panacea in the time of COVID-19 crisis. *Journal of Educational Technology Systems*, 49(1), 5–22.

Dincol, S., Temel, S., Oskay, O. O., Erdogan, U. I., & Yilmaz, A. (2011). The effect of matching learning styles with teaching styles on success. *Procedia - Social and Behavioral Sciences*, 15(1), 854–858. https://doi.org/10.1016/j.sbspro.2011.03.198

Entwistle, N. (2000). *Promoting Deep Learning through Teaching and Assessment: Conceptual Frameworks and Educational Contexts*. Leicester: TLRP Conference.

Gloria, R. & Benjamin, A. E. W. (2014). Techno-pedagogical skills in teacher education. *International Journal of Scientific Research*, 12(3), 91–92.

Hiltz, R. & Turoff, M. (1978). *The Network Nation: Human Communication via Computer*. Reading, MA: Addison-Wesley Infusion of Techno-Pedagogy during Covid-19: Teachers' perspective SkMonirul Islam 2020 IJCRT | 8(11),| ISSN: 2320-2882 IJCRT2011433 International Journal of Creative Research Thoughts (IJCRT) www.ijcrt.org 3660-3670

Khandaghi, M. A. & Rajaei, M. (2011). The comparison of students and educators preferred teaching styles in teacher educating centers. *Procedia - Social and Behavioral Sciences*, 15(1), 1875–1880. https://doi.org/10.1016/j.sbspro.2011.04.019

Koushlesh, M. & Yadav, P. (2015). Modern trends in teaching. In A. Pathak, & S. Francis (Eds.), *TechnoPedagogy and Learning* (pp. 1–18). Lucknow: Vedant Publications.

Kumari, V. (2008). A study of thinking and teaching styles of teacher educators about some selected demographic variables (Doctoral Dissertation, Himachal Pradesh University). Retrieved from http://hdl.handle.net/10603/129145

Laurillard, D. (2001). *Rethinking University Teaching: A Conversational Framework for the Effective Use of Learning Technologies*. New York/London: Routledge.

Marton, F. & Saljö, R. (1997). Approaches to learning. In F. Marton, D. Hounsell, & N. Entwistle (Eds.), *The Experience of Learning*. Edinburgh: Scottish Academic Press (out of the press, but available online).

National Education Policy (2020). https://www.education.gov.in/ referred on 20/08/2020.

Pask, G. (1975). *Conversation, Cognition, and Learning*. Amsterdam/London: Elsevier (out of the press, but available online here)

Saranyadevi, S. (December 2018). Influence of emotional intelligence on professional commitment of high school teachers. *Indian Journal of Research*, 7(12), 33–34. https://www.worldwidejournals.com/paripex/recent_issues_pdf/2018/December/December_2018_1544015119__92.pdf

Scardamalia, M. & Bereiter, C. (2006). Knowledge building: Theory, pedagogy, and technology. In K. Sawyer (Ed.), *Cambridge Handbook of the Learning Sciences* (pp. 97–118). New York: Cambridge University Press.

Shaari, A. S., Yusoff, N. M., Ghazali, I. M., Osman, R. H., & Dzahir, N. F. M. (February 2014). The relationship between lecturers' teaching style and students' academic engagement. *Procedia - Social and Behavioral Sciences*, 118(1), 10–20. http://doi:10.1016/j.sbspro.2014.02.002

Shah, N. (2014). The future of classroom: The role of teachers needs a relook in the digital era. Retrieved from http://indianexpress.com/article/lifestyle/the-future-classroom-the-roleof-teachers-needs-a-relook-in-digital-era/99/print/

Sondhiya, R. (2015). Role of the teacher in ICT pedagogy. In A. Pathak & S. Francis (Eds.), *Techno-Pedagogy and Learning* (pp. 150–159).Lucknow: Vedant Publications www.ijcrt.org © 2020 IJCRT |8 (11), November 2020 | ISSN: 2320-2882 IJCRT2011433 International Journal of Creative Research Thoughts (IJCRT) www.ijcrt.org 3670

Thakur, N. (2015). A study on the implementation of techno-pedagogical skills, its challenges, and role to release at a higher level of education. *American International Journal of Research in Humanities, Arts and Social Sciences*, 9(2), 182–186.

Vajargah, K. F., Jahani, S., & Azadmanesh, N. (2010). Application of ICTs in teaching and learning at university level: the case of ShahidBeheshti University. *The Turkish Online Journal of Educational Technology*, 9(2), 33–39.

Vijaya, R. (2017). Techno-pedagogy in teaching and learning. *Shanlax International Journal of Education*, 5, ISSN: 2320-2653.

Yildiz, A. (2017). The factors affecting techno-pedagogical competencies and critical thinking skills of preservice mathematics teachers. *MOJES: Malaysian Online Journal of Educational Sciences*, 5(2), 66–81.

Chapter 12

Digital Evaluation Tools

Jijish Elias and M. Mirunalini

Digital Evaluation in the Context of the Pandemic

Evaluation is an unavoidable process in teaching-learning activities. Reviewing assignments, evaluating them, and grading is the usual routine of a normal teacher. Not only assignments but a lot of other activities should be evaluated by the teacher in the academic environment. The development of mobile technology, web-based evaluation tools, and online evaluation techniques make it easy and quick for teachers (India Today, 2020).

All have recognized the necessity to go in hand with technology during the pandemic. COVID-19 has, nevertheless, strengthened our economy and has raised significant intergenerational issues. Throughout this period, meanwhile, investigations, and virtual tests show several advantages of automated evaluation. It saves educators adequate time and saves students the hassle of presenting their expertise in the various fields of their disciplines.

Digital Evaluation Tools for Teachers

The usage and development of technologies applicable to educational purposes are increasing day by day. Teachers and students are getting more adapted to such technologies. While considering the evaluation of academic achievement, the process of examination is inevitable, but classical paper-based testing is losing its position in the academic process, and online tools are coming up in the field.

Incorporation of learning management systems and online feedback techniques like Google Forms and Survey Monkey are applicable in the evaluation

 DOI: 10.1201/9781003504894-12

process, which provides fast and fair reports of assessment immediately after the assessment.

Features of Digital Evaluation Tools

Various features and characteristics of digital evaluation tools experimented with by the academic community are discussed below (Johnson, 2008; Woline, 2009). These characteristics may differ in one or another sense from one tool to another (Paiva et al., 2020).

Time-Saving

One major attraction of digital evaluation tools is their time-saving character (Darwin, 2016).

Very little duration is required to execute the evaluation process for plenty of target populations. Once the evaluation tool is prepared, it can be utilized many times with small changes or without any changes. For example, the unit tests prepared for a batch can be utilized for successive batches without any changes until the syllabus is the same. Similarly, most digital evaluation tools give provisions to combine questions from different question papers. In such cases, the tool can be utilized with small changes. For example, a terminal examination can be prepared by combining the selected questions from each chapter which are considered for evaluation.

Comprehensive

The comprehensive nature of digital evaluation tools is remarkable. The manual evaluation process is always connected with bias in the assessment. Digital evaluation tools overcome this issue, and they focus on the practicality of the topic covered in the evaluation process rather than checking the theoretical knowledge. Subjective-type questions are avoided in most of the digital evaluation tools and objectivity of the evaluation process is kept.

Student-Friendly

Most of the evaluation tools in digital media are self-explanatory and student-friendly. Access to the tools by the candidate will be made easy and friendly in digital evaluation tools.

Quick Feedback

Most digital evaluation tools provide immediate feedback after the evaluation process. Most of the tools contain a feedback system with explanations for the wrong answers made by the candidate. Even though there are digital tools that at video feedback after the manual verification of instructors too.

Authentic Monitoring

Some of the online or digital evaluation tools do not allow candidates to move to other applications or browsers. Some tools use video capturing systems and in such cases, even the movement of the student, movement of the lips, and voice detection facilities are utilized to prevent malpractices.

Punctuality

Punctuality is another important characteristic of a digital evaluation tool. The examiner can fix a particular time to start and stop the evaluation process. The candidates will get indications or alarms during the evaluation process regarding the duration left to finish with the help of a timer. There are examinations and evaluation systems available in digital form which allows the candidate to take their own time.

Versatility

The versatile nature of digital evaluation tools also makes them more attractive. Students can select different levels of examination patterns according to their subject knowledge and move to the next level at their own pace of learning. Availability to practice examinations in offline mode is also given by the digital evaluation tools. A student can take a needed number of practices without any extra cost using digital evaluation tools.

Types of Digital Evaluation Tools

The following are the major types of digital evaluation tools (eLearning Industry, 2021).

1. **Gamifying the answers**
   Evaluation tools, like Kahoot!, Answer Garden, etc., make the evaluation process into a game. The students should discuss collaboratively to complete their tasks. Students are engaged with the process during the assessment. Mostly, multiple-choice questions are prepared by the teacher or available in the tool used in this type of evaluation, but they are just like in a game.
2. **Survey-based assessment**
   Digital tools, like Google Forms, Survey Monkey, etc., make use of methods of surveying in the evaluation process. Several kinds of questions, like

multiple choice, fill in the blank, match the following, etc., are available in this kind of tool.

3. **Tracking the progress**

   Evaluation tools like Plickers assess the students by tracking the level of progress they have reached. The tool tracks the achievement of students at each level they have reached by completing the tasks of the course enrolled by the student.

Popular Digital Evaluation Interfaces in India

A few digital evaluation interfaces that are commonly used in the Indian scenario are discussed below.

1. **Google form**

   It is the most common interface used by teachers and other academic agencies for collecting feedback and conducting online quizzes. It is very easy to prepare, conduct, and analyze the results. Several types of questions and images can be added to this tool to customize the product.
2. **Survey monkey**

   Survey monkey also has an interface that is similar to Google Forms. Even though it can be used as an evaluation tool, it is still more famous as a data collection tool.
3. **Survey heart**

   This is also similar to both discussed above and is more smartphone-friendly. One important disadvantage of Google Forms is that it is not possible to use it as an applet on a smartphone.
4. **eXe**

   eXe is a software available as an open resource. It helps to create interactive evaluation tools that can be utilized both offline and online.
5. **Hot potatoes**

   Hot potatoes is also an offline evaluation tool maker software. The teachers can make interactive type evaluation tools with hot potatoes.
6. **Quiz**

   It is an interesting online quiz maker. This tool is used to interact with students during online sessions by academicians. Listeners can make polls and attend quizzes while they attend a webinar to get to interact with the presenter.
7. **Slido**

   This is an add-on available in Google Presentation. This helps to make the Google presentation platform into an evaluation platform. Students can attend the quiz or poll using Slido when the evaluator shares it.

Preparation of Candidates for Online Examinations

When considering the Indian scenario, online digital evaluation is not common. Students may follow the following guidelines to prepare for their online evaluations (India today, 2021; Latest Education Technology News, 2019).

1. **Attend mock tests**
   Each candidate should attend a maximum number of mock tests before attending the real test. This will help them to get familiarized with the interface of the evaluation tool and the options available in the evaluation tool. It helps to keep the time limit for the examination.
2. **Check internet connectivity**
   Before attending the real examinations, the candidate should make sure that there will be uninterrupted internet connectivity and power supply. Along with this, confirm that the device that is going to be used during the evaluation will be adaptable to the interface of the evaluation tool.
3. **Go through dos and don'ts**
   All evaluation systems will be connected with some dos and don'ts, which should be strictly followed. It is the responsibility of each candidate to get familiarized with the dos and don'ts of the evaluation going to attend. Students should be well aware of the invigilation process and anti-cheating components available in the evaluation interface.

To Sum Up

As the pandemic situation compelled the learning as the evaluation process digital mode, the usage of adequate technologies and interfaces by teachers and students became inevitable for the academic community. Most of the educational systems and boards tried their turn to conduct the evaluation process in digital mode and some made it happen while others failed. In this context, it is very much required by the academic community to get aware of digital evaluation tools and how to make them useful.

References

COVID-19 impact: Is online learning the new future of education? (2020, June 30). India Today. https://www.indiatoday.in/education-today/featurephilia/story/ covid-19-impact-is-online-learning-the-new-future-of-education-1690483-2020-06-30

Darwin, S. (2016). Charting new approaches to student evaluation. *Student Evaluation in Higher Education*, 151–173. https://doi.org/10.1007/978-3-319-41893-3_9

Digital assessment tools: What is their impact on a teacher's work? (2021, May 12). eLearning Industry. https://elearningindustry.com/digital-assessment-tools-impact-teachers-work

How to prepare students for online exams. (2021, June 18). India Today. https://www.india-today.in/education-today/featurephilia/story/how-to-prepare-students-for-online-exams-1816662-2021-06-18

Indian Edtech in 2019 – Part 1. (2019, February 12). Latest Education Technology News. https://techplusmedia.com/education/2019/02/12/indian-edtech-in-2019-part-1/

Johnson, G. M. (2008). Student evaluation of online study tools questionnaire. *PsycTESTSDataset.* https://doi.org/10.1037/t13455-000

Paiva, S., Oliveira, L. R., & Reis, L. P. (2020).An assessment classification of formative evaluation tools and digital tools. *ICERI2020 Proceedings.* https://doi.org/10.21125/iceri.2020.0730

Woline, T. (2009). Evaluation of online educational tools in horticulture: An online course and problem-solving software. https://doi.org/10.31274/etd-180810-2097

Chapter 13

Student Assessment and Evaluation in the Disruptive Model of Teaching and Learning in Indian Higher Education

A Critical Analysis

Priya Topno and Parimal Sarkar

Objectives of the Chapter

The chapter aims to fulfill the following objectives which are general and specific.

- **General Objectives:**
  - To critically examine and analyze the issue of student assessment and evaluation in the disruptive model of teaching and learning in higher education in the Indian context.
- **Specific Objectives:**
  - To discuss the concept of the disruptive model in education both from theoretical and applied perspectives.
  - To explain the concept of assessment and evaluation in education.
  - To traverse through assessment and evaluation in the disruptive model, to explore its advantages and challenges, and to glance through the policy

DOI: 10.1201/9781003504894-13

perspective of digital assessment and evaluation in higher education in India.

- To recommend ways to ameliorate the gaps and differences in implementing digital assessment and evaluation in higher education in India.

Background

In every sense of the word, education is a purposeful activity that strives to achieve its goals through various endeavors. One of the primary goals of education has always been in favor of the interest and welfare of the student. There have been multifarious efforts for centuries on the part of various stakeholders of education to make the educational process an effective one through which the student can gain maximum benefit and contribute to society further. To validate and ensure the efficacy of this virtuously interdependent relationship of student development and contribution, the important and integral dimensions of education are to be considered, and they are known as student assessment and evaluation.

Student assessment and evaluation are all-encompassing terms that appraise and ascertain the achievements and outcomes of the learners in various aspects of the educational process undertaken by them. Student assessment and evaluation practices are pertinent to a formalized or organized and structured educational system and, therefore, are carried out at different intervals of a learner's educational journey. With the focus of education, in general, turning more toward the development of the whole personality of the learner, the trends in student assessment and evaluation have also undergone a major overhaul. The age-old examination, which the student had to undergo at the end of the academic term, is evolving and is gradually being replaced by more ingenious and effective ways of determining student progress with each passing day as changes and development take place in society.

One inevitable and massive change that society has witnessed in recent years is the change brought about by technological advancement and its invasion into the lives of the common people. Education is no exception when it comes to the impact that technology has had on humans. In keeping with the dynamic nature of education, it has yet again adapted itself to the changing contours of the given times by adopting technological advancements and bringing about significant ramifications in student assessment and evaluation. The World Wide Web was introduced in the 1990s and since then many companies have introduced their e-assessment systems (Alruwais, Wills, and Wald, 2018).

This chapter unveils the very relevant issue of the disruptive strategy of teaching and learning, which has taken over the education system throughout the world in this age of technology, and puts into perspective its various facets with a focus on student assessment and evaluation in the context of higher education in India.

Concept of Disruptive Model

Theoretical Framework

The meaning of disruption, as stated in Oxford Learner's Dictionaries, is "a situation in which it is difficult for something to continue in the normal way; the act of stopping something from continuing in the normal way." The term disruptive, being an adjective of the term disruption, carries the same meaning. Hence, the disruptive model is that which intends to break with the established model to improve the existing one. The disruptive model in education has its origins in Bower and Christensen's (1995) disruptive innovation theory which focuses on disruptive technologies. Christensen, Raynor, and McDonald (2015) clarify that "classic disruption entails—a small enterprise targeting overlooked customers with a novel but modest offering and gradually moving upmarket to challenge the industry leaders." The theory though primarily concerned with technological innovations and economic growth holds tremendous potential in educational and social welfare as well. The theory of disruption can provide researchers, practitioners, and policymakers with a new perspective on increasingly affordable and accessible educational opportunities in our society (Christensen, Aaron and Clark, 2003, p. 45). Rapid progress in educational practices and processes aided by technology has resulted in multiple possibilities to customize education by creating alternative possibilities to access education to suit and cater to the learning requirements of the individual as well as the requisites of society. The disruptive model of education facilitates the digitalization of education by moving away from and revamping the previous model of education; and many experts consider this change to be necessary and urgent to transform the existing system, still anchored in the previous century, to address the needs of the digital age.

Applied Aspect

The document titled *Evaluation Reforms in Higher Educational Institutions*, 2019 put forth by the University Grants Commission (UGC) mentions various alternatives and disruptions in the traditional system of education with the potential to augment and revolutionize learning as well as assessment and evaluation methods. Some of the disruptive adaptations suggested in the document can be listed as synchronous learning, semi-synchronous learning, blended learning, collaborative learning, flipped classroom, and so on. To facilitate lifelong learning technology-based platforms like Massive Open Online Courses (MOOCs) especially provided through Study Webs of Active–Learning for Young Aspiring Minds (SWAYAM) can be utilized. Another alternative is the learning management system (LMS), which can be used by institutions to integrate the entire teaching, learning, and evaluation process. The LMS if used appropriately can aid higher educational institutions to both deliver academic content in blended form as well as to assess learning through in-built activities like thesis, assignments, workshops, discussion forums, and quizzes. Some of the open-source LMSs such as Moodle and Edmodo provide a variety of ways to post content in the form of videos, audio, e-learning modules, live class sessions, and so on.

A study conducted by Gaytan and McEwen (2007) shows that

> effective assessment techniques, as perceived by faculty and student respondents, include projects, portfolios, self-assessments, peer evaluations, weekly assignments with immediate feedback, timed tests and quizzes, and asynchronous type of communication using the discussion board. The use of rubrics to aid the assessments and to provide meaningful and quick feedback is valued by instructors and students alike.
>
> *(p. 129)*

Concept of Assessment and Evaluation

Assessment and evaluation are two such components of the teaching-learning process that with their absence the entire exercise would remain incomplete. The aims, objectives, and purposes with which any educational journey begins, find their culmination and fulfillment through the process of assessment and evaluation. The two terms are so closely related to one another in their purpose that they are more often used interchangeably than exclusively. Assessment and evaluation as observed by Stobart (2017)

> play an ever-increasing role in modern society. They often determine the life chances of those assessed, shaping learner identities. These are tests which are deliberately designed to gather specific information about individuals so a judgment can be made, for example, about whether they should be selected for progression to a particular institution or be allowed to practice in an occupation.
>
> *(p. 56)*

On one hand, though assessment and evaluation are crucial indicators of the efficacy of the teaching-learning process, on the other hand, they are ever challenged with the task of keeping up pace with the rapid changes and complexities of the educational process. There, however, lies a subtle difference between the two in their orientation of achievement. This difference is evident in the fact that, while assessment is process-oriented, evaluation is product-oriented. In other words, assessment is a process of collecting, reviewing, and using data, for improvement in the current performance while evaluation is described as an act of passing judgment based on a set of standards. Brown (2004) defined assessment as "any act of interpreting information about student performance, collected through any of a multitude of a means or practices" (p. 304). Evaluation refers to the process of arriving at judgments about abstract entities such as programs, curricula, organizations, institutions, and individuals (Ghaicha, 2016, p. 213).

The assessment and evaluation procedures are of four types – formative, summative, diagnostic, and integrative (Crisp, 2011). The credit for the coinage of the terms, formative and summative evaluation goes to Scriven (1967). He stressed their disparities in terms of the information they seek and how it is used, as well as the purposes of the information they seek. Later, Bloom (1968) adopted the term in his book *Learning for Mastery* to consider formative assessment as a tool for improving the teaching-learning process for students. According to Ghouali, Benmoussat, and Ruiz-Cecilia (2020), formative assessment is a source of data through which instructors highlight the difficulties faced by their learners and make appropriate changes to refine their teaching, and learners identify their deficiencies and improve their learning abilities. While summative assessment seeks the measurement of the quantity of the student's knowledge once the instruction is finished as it is interested mainly in the learners' or the program's results (Hyland, 2003, as cited in Ghouali, Benmoussat, and Ruiz-Cecilia, 2020, p. 53). Diagnostic assessment is often the most underused of the traditional assessment formats in tertiary education and is too often associated with a negative model designed to identify deficiencies in students` capabilities (Benseman and Sutton, 2008). Integrative assessments are designed to promote and measure student self-regulation and the capabilities associated with lifelong learning (Crisp, 2011).

The Traditional Practice of Assessment and Evaluation

The traditional ways of assessment and evaluation have been as per the traditional methods of instruction, teaching, and learning. However, in a relatively new era of technology-based teaching and learning, assessment and evaluation practices require a more ongoing, systematic approach than that used with traditional instruction (Robles and Braathen, 2002). Crisp (2011) and Jordan (2013) agree that the traditional methods of assessment and evaluation lack immediate and direct feedback, students' involvement, and over-emphasis on grades resulting in a lack of motivation, creativity, and self-confidence among the learners. Stating the international scenario about the conventional practices of assessment and evaluation, the UGC in the document titled *Evaluation Reforms in Higher Education Institutions*, 2019 cites the report of a quality audit of 23 universities by the United Nations Development Programme Regional Bureau for the Arab States as;

> assessment continues to be a weakness, particularly in three respects. First, too much emphasis is placed on the memory recall of descriptive knowledge. Second, not enough is done to test higher-level cognitive

> skills. Third, there is virtually no moderation either internal or external to ensure the fairness and transparency of marking.
>
> *(p. 12)*

This resonates to a great extent with the current plight of assessment and evaluation in Indian higher education institutions by and large. The present system tests memory learning skills which may be required to a certain extent but are not adequate to equip the learner for the challenging environment that currently prevails. Further, standardizing assessment and evaluation is analogous to "one size fits all" which undermines the importance of individual differences and abilities. With the given scenario, it is, therefore, imperative that assessment and evaluation modes be switched from the single vertical mode to multiple horizontal modes to ensure the credibility and efficacy of the assessment and evaluation system. In addition, the evaluation and assessment methods must match the level of desired competencies, and so require educators to modify their methods of instruction to make them more innovative than traditional instruction (Liang and Creasy, 2004).

Assessment and Evaluation in the Disruptive Model

Assessment and evaluation in the disruptive model of teaching and learning are technology-based and draw heavily from the three categories stated in disruptive innovation theory. The three categories are (a) technologies that enable us to do something we had already been doing a little bit better than before, (b) technologies that prompt new practices, and (c) technologies that make things possible to do more with less (Flavin and Quintero, 2018). Digital assessment and evaluation, therefore, enable us to assess and evaluate a little better than before, prompt new practices, and help achieve more with less. "The technological enhancement that has been witnessed in today's world has taken on new perspectives towards assessment (Jordan, 2013) that have diversified its formats and tools and led to a technology-based called e-assessment" (Mojarrad *et al.*, 2014, as cited in Ghouali, Benmoussat, and Ruiz-Cecilia, 2020). Crisp (2011) puts forward a more detailed explanation and claims that in this ICT (Information and Communication Technology)-based evaluation, students' assignments, answers, scores, and feedback are elaborated, distributed, recorded, corrected, analyzed; stored via digital devices that can be computers, mobile phones, tablets, or gaming tools, and presented in various layouts such as texts, documents, sounds, pictures, videos, or games (Ghouali, Benmoussat and Ruiz-Cecilia, 2020).

Blended Mode of Teaching and Learning: Concept Note, 2021 proposed by the UGC suggests implementing blended learning as a new mode of teaching-learning in higher education, a disruptive measure to the existing one, and hence necessitates the area of assessment and evaluation to be explored in the light of blended learning.

It suggests some important strategies like continuous comprehensive evaluation, innovation in formative and summative evaluation through open-book examinations, group examinations or e-portfolios, and creative products, to name a few, incorporation of assessment rubrics stating the level of performance and descriptors to be integral aspects of assessment and evaluation in the disruptive model of teaching and learning in higher education in India.

Most of the studies agree that in the disruptive model, assessment and evaluation procedures from the start to the end of the assessment should be carried out electronically. This means that the design, test implementation, recording of the response, and providing feedback are all completed using ICT (Alruwais, Wills, and Wald, 2018).

Whitelock *et al.* (2006) clarify how e-assessment can work by providing cycles of the process (Figure 13.1). They claim that this framework enables educationalists to address the barriers and the cultural debate surrounding e-assessment strategies (Alruwais, Wills and Wald, 2018, p. 34). The e-assessment cycle suggests that motivation is the first and crucial stage in the disruptive model of assessment and evaluation. It is followed by the design and creation of the assessment. Then the test is administered to the students and after they finish the outcomes are delivered. The data is then processed and feedback is provided. Finally, an evaluation of the outcomes and a review of the feedback is done. After this last step, the cycle moves back to the design and creation steps; thus, the testing, data retrieval, or evaluation outcomes, depending on the outcomes and feedback to enhance the assessment and feedback to meet the aims of the

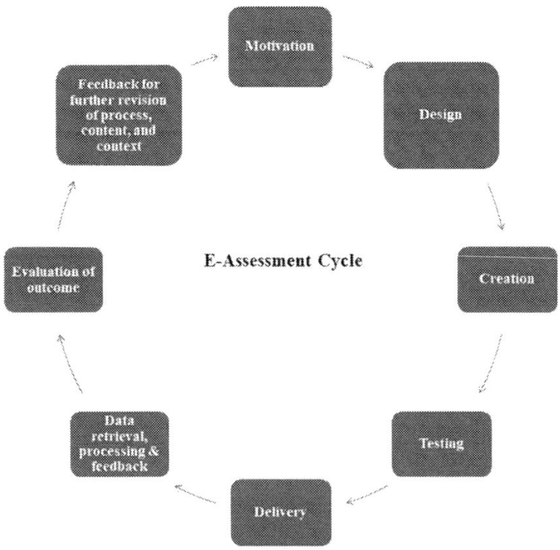

Figure 13.1 e-Assessment Cycle.

course and reach the desired outcomes. These stages of assessment distinguish the disruptive model of assessment and evaluation from the traditional system of assessment and evaluation.

Advantages

The technological intervention in the area of assessment and evaluation reduces dependency on human interference and minimizes the chances of errors. Electronic assessment has the potential to help students become more reflective learners. This is largely due to the speed with which feedback can be provided (Whitelock, 2006). Some of the advantages in this regard are listed below:

i. Automation of the entire assessment and evaluation procedure right from registration to convocation.
ii. Objectivity in judging student performance.
iii. Transparency in the marking system.
iv. Economy of time.
v. Tracking students' performance.
vi. Student involvement in the assessment and evaluation procedure.
vii. Digitization of answer scripts and their on-screen evaluation.
viii. Systematic and effective documentation of records.

Challenges

The advantages and benefits of the digital assessment and evaluation prompted by disruptive technology add to the value and relevance of the present-day education system, and efforts are on from the various stakeholders to keep pace with the changing times. However, with the diversity and complexity of various kinds in India, the challenges surrounding digital assessment and evaluation are many. Different studies, Donovan, Mader and Shinsky (2007), Crews and Curtis (2010), and Thadathil *et al.* (2020) have examined the challenges in technology-based education and some of the road-blocks in implementing digital assessment and evaluation can be stated below:

i. Vast digital divide in terms of access to the digital world.
ii. Digital illiteracy in terms of usage.
iii. Insufficient digital infrastructure.
iv. Network and power disruptions.
v. Inadequately trained teachers to handle the digital interface of assessment and evaluation.
vi. Apprehension and lack of awareness about digital assessment and evaluation.
vii. Limited human interface affecting interpersonal relationships.

Digital Assessment and Evaluation in Higher Education in India: Policy Perspective

The disruptive model of teaching and learning provides a flexible structure to the educational process and opens up multiple avenues in terms of accessibility and student assessment and evaluation. One, of the multiple opportunities that the disruptive model lends, is the student assessment and evaluation in the digital form. In the Indian context, the National Education Policy, 2020 endorses the leverage of technology for teaching-learning at all levels from school to higher education. In this context, about digital assessment and evaluation, the policy states;

> Appropriate bodies, such as the proposed National Assessment Centre or PARAKH, School Boards, NTA, and other identified bodies will design and implement assessment frameworks encompassing design of competencies, portfolio, rubrics, standardized assessments, and assessment analytics. Studies will be undertaken to pilot new ways of assessment using education technologies focusing on 21st-century skills.
>
> *(NEP, 2020 p. 60)*

The pace at which technology is evolving, and its inclusion in education is an ongoing journey rather than a destination to be reached. Hence, adapting to technology and adopting technology to education is a continuous process and results in numerous efforts with trials and errors. What works today may turn out to be irrelevant in a couple of years which leaves the stakeholders of education with various advantages to explore and challenges to overcome to achieve the educational goal of developing the individual and the society.

Recommendations

Technology and its benefits are aplenty in the current scenario and to reap these benefits, it is important to bridge the gap and resolve the challenges that lie in implementing technological innovations and advancement to maximize educational returns in the 21st century. With this end in view, the following recommendations have been laid down, keeping the National Education Policy 2020 initiatives as the basis.

i. Identify and categorize emergent technologies based on their potential and estimated time frame for disruption.
ii. Expand research efforts in disruptive technology.
iii. Strengthen assessment and evaluation in the disruptive model by using a wide variety of educational software and by seeking assistance from specialized and appropriate bodies like National Assessment Centre, National Testing Agency, and so on.

iv. Encourage systematic and advanced forms of assessment and evaluation based on rubrics to trace the achievement of teaching objectives and learning outcomes.
v. Initiate standardized automated evaluation and assessment which are user-friendly and inclusive.
vi. Introduce blended modes of learning, MOOCs, LMS, etc., as measures of disruptive educational strategies.
vii. Provide training and incentives for teachers to equip them with pedagogies and techniques related to digital education, assessment, and evaluation.
viii. Address the digital divide by increasing digital accessibility.
ix. Improve digital infrastructure to solve India's challenges in terms of scale, diversity, complexity, and device penetration.
x. Support research and pilot studies for online education, the results of which can be used for continuous improvement.

Conclusion

Education is again at the crossroads with emerging technologies making their way into our everyday lives and impacting educational practices, processes, and products enormously. The disruption caused by technology in every aspect of human life has caused an inevitable ripple effect in the realm of education as well. The disruptive model of teaching and learning is fast catching up and replacing traditional and existing models of education, ensuing changes, modifications, and adaptations to cater to the demands of the 21st-century requirements.

Assessment and evaluation being indispensable components of education are required to align with the changing scenario, and parameters of testing and assessment need to be relooked and reoriented to cater to the requirements of the present generation of learners to provide a wholesome educational experience. An overhaul of assessment and evaluation tools using digital platforms needs to be put in place, which would suffice for the requisite skills to be instilled in the learners of a growing economy and meet the demands of the 21st-century society. This can be materialized through unified efforts on the part of various stakeholders of education because the progress of any nation or society depends greatly on its education system and how it advances.

References

Alruwais, N., Wills, G., & Wald, M. (2018). Advantages and Challenges of Using e-Assessment. *International Journal of Information and Education Technology, 18*, 34–37. Retrieved February 1, 2022, from http://www.ijiet.org/vol8/1008-JR261.pdf

Benseman, J., & Sutton, A. (2008). *OECD/CERI Formative Assessment Project background report: New Zealand.* Retrieved November 10, 2021, from https://www.oecd.org/newzealand/40015787.pdf

Bloom, B. (1968). Learning for Mastery. *Evaluation Comment.* Retrieved February 3, 2022, from https://programs.honolulu.hawaii.edu/intranet/sites/programs.honolulu.hawaii.edu.intranet/files/upstf-student-success-bloom-1968.pdf

Bower, J. L., & Christensen, C. M. (1995). Disruptive Technologies: Catching the Wave. *Journal of Product Innovation Management.* https://doi.org/10.1016/0024-6301(95)91075-1

Brown, G. (2004). Teachers' Conceptions of Assessment: Implications for Policy and Professional Development. *Assessment in Education Principles Policy and Practice, 11*(3), 301–318. https://doi.org/10.1080/0969594042000304609

Christensen, C. M., Aaron, S., & Clark, W. (2003, January). *Disruption in Education.* Retrieved August 2021, from Educase Review: https://er.educause.edu/-/media/files/articles/2007/1/erm0313.pdf?la=en&hash=F36B08C108E63EE7321FFA4DA64E8E53471B9EAD

Christensen, C. M., Raynor, M. E., & McDonald, R. (2015, December). *Harvard Business Review.* Retrieved August 27, 2021, from Harvard Business Review Web site: https://hbr.org/2015/12/what-is-disruptive-innovation

Crews, T. B., & Curtis, D. F. (2010). Online Course Evaluations: Faculty Perspective and Strategies for Improved Response Rates. *Assessment &Evaluation in Higher Education,* 865–878. https://doi.org/10.1080/02602938.2010.493970

Crisp, G. (2011). *Teacher's Handbook on e-Assessment.* Retrieved February 4, 2022, from http://transformingassessment.com/sites/default/files/files/Handbook_for_teachers.pdf

Donovan, J., Mader, C., & Shinsky, J. (2007). Online vs. Traditional Course Evaluation Formats: Student Perceptions. *Journal of Interactive Online Learning, 6*(3), 158–180. Retrieved February 2, 2022, from http://www.ncolr.org/jiol/issues/pdf/6.3.2.pdf

Flavin, M., & Quintero, V. (2018). UK Higher Education Institutions' Technology-Enhanced Learning Strategies from the Perspective of Disruptive Innovation. *Research in Learning Technology,* 1–12. https://doi.org/10.25304/rlt.v26.1987

Gaytan, J., & McEwen, B. C. (2007). Effective Online Instructional and Assessment Strategies. *American Journal of Distance Education,* 117–132. https://doi.org/10.1080/08923640701341653

Ghaicha, A. (2016). Theoretical Framework for Educational Assessment: A Synoptic Review. *Journal of Education and Practice, 7*(24), 212–231. Retrieved August 2021, from https://eric.ed.gov/?id=EJ1112912

Ghouali, K., Benmoussat, S., & Ruiz-Cecilia, R. (2020, January 25). E-assessment on the spotlight: Present and prospects. *Reidocrea,* 52–62. https://doi.org/10.30827/Digibug.59151

Government of India. (2019). *Evaluation Reforms in Higher Educational Institutions.* New Delhi: Secretary, University Grants Commission. Retrieved from https://www.ugc.ac.in/ebook.aspx

Hyland, K. (2003). *Second Language Writing.* Cambridge University Press.

Jordan, S. (2013). E-assessment: Past, Present and Future. *New Directions in the Teaching of Physical Sciences,* (9), 87–106. https://doi.org/10.29311/ndtps.v0i9.504

Liang, X., & Creasy, K. (2004). Classroom Assessment in Web-Based Instructional Environment: Instructors' Experience. *Practical Assessment, Research & Evaluation, 9*(7), 1–9. https://doi.org/10.7275/84mr-wp41.

Ministry of Human Resource Development. (2020). *National Education Policy 2020.* Government of India.

Mojarrad et al. (2014). Computer-based Assessment Vs. Paper Pencil-based Assessment. *An Investigation into the Performance and Attitude of Iranian EFL Learners' Reading Comprehension. International Journal of Language Learning and Applied Linguistics World, 4*, 2289–2737.

Robles, M., & Braathen, S. (2002). Online Assessment Techniques. *The Delta Pi Epsilon Journal, 44*(1), 39–49.

Scriven, M. (1967). The Methodology of Evaluation. *Social Science Education Consortium.* Retrieved February 3, 2022, from https://www.scribd.com/doc/283341599/Scriven-The-Methodology-of-Evaluation-1967

Stobart, G. (2017). Assessment and Learner Identity. In M. A. Peters (Ed.), *Encyclopedia of Educational Philosophy and Theory* (pp. 56–60). Springer.

Thadathil, G., Chambi, W., Prasad, Y., & Silva, E. (2020). The Salesian College of India and The Salesian University of Bolivia in the Context of the Pandemic. *Estudios Pedagogicos, 46*(3), 287–301. Retrieved from https://pesquisa.bvsalud.org/global-literature-on-novel-coronavirus-2019-ncov/resource/pt/covidwho-1090258?lang=en

University Grants Commission. (2021). *University Grants Commission.* Retrieved from University Grants Commission Website: https://www.ugc.ac.in/searchresult.aspx?q=Concept+Note+on+Blended+Learning&cx=000457167406620127635%3afugsadtztkk&cof=FORID%3a9

Whitelock, D. (2006). Electronic Assessment: Marking, Monitoring and Mediating Learning. *International Journal of Learning Technology, 2*(2-3), 264–276. https://doi.org/10.1504/IJLT.2006.010620

Whitelock, D., Reudel, C., & Mackenzie, D. (2006). Identifying innovative and effective practice in e-assessment: findings from seventeen UK case studies. In: Danson, Myles ed. *Proceedings of the 10th CAA International Computer Assisted Assessment Conference* (pp. 505–511). Loughborough: Professional Development, Loughborough University.

Chapter 14

Teacher Capacity Building in High-Performing Digital Education Ecosystem

Vu Thi Lan Anh, Dinh Thi Phuong Hoa, Nguyen Thi Linh Yen, and Ton Quang Cuong

Introduction

The process of digitization and equalization in access to teaching on today's digital platforms is currently a dominant trend, strongly promoting the production and acquisition of knowledge and educational models. The digital transformation simultaneously creates "non-traditional" teaching methods. Educational institutions are no longer simply defined as limitedly closed spatial contexts (including campuses, classrooms, compulsory curricula, and one single-textbook system). Multimedia learning methods and advanced educational technologies are now closely linked with the process of teaching and learning, whether formal, non-formal, or informal. This transformation serves individual demands for education, promoting personalized learning experiences for each learner. The trend poses a challenge to training and professional competence development for both pre-service and in-service teachers.

The national project to enhance the capacity of lecturers and managers of higher education institutions to meet the requirements of fundamental innovation, comprehensive education, and training phases of 2019–2030 stated the priority of

 DOI: 10.1201/9781003504894-14

building a team of faculty and staff to manage higher education institutions to ensure quality, rational structure, and meet the fundamental innovation requirements. Decision No. 89/QD-TTg (2019) emphasizes national educational tasks in the period 2019–2030 to "comprehensive education, train high-quality human resources, and develop science and technology for the country, attached to the requirements of entrepreneurship, innovation, and the demands of the 4.0 industrial revolution."

In the context of the COVID-19 pandemic, integrating technology solutions and adapting them to teaching and teachers' competence development has become more urgent than ever. In this situation, "education has become an emergency matter, and along with it, educational technologies have been positioned as a frontline emergency service," or "the possible contraction of higher education as an on-campus experience and a shift to remote instruction and learning is already concerning many educators" (Ben Williamson, Rebecca Eynon, and John Potter, 2020).

With the increased importance of developing new "pandemic pedagogy" to avoid "the traumas inflicted by the pandemic," Roy Schwartzman (2020) stated, "more vigorous practice of communal, care-based, collaborative resilience through reimagining the nature and purpose of communication instruction... digital divides based on disparities in access, skill, and technological features; and the reassertion of neoliberal approaches to education." Furthermore, COVID-19 refers to the process of learning across various settings through social and content exchanges, with different perspectives toward online education globally due to the large-scale use of online platforms by learners and educators (Shahnawaz Saikat, Jaspaljeet Singh Dhillon, Wan Fatimah Wan Ahmad, and Robiatul A'dawiah Jamaluddin, 2021).

At the same time, it is necessary to design and build a general model of this process to refer to and integrate into the adjustments of the content and implementing methods of educational programs on a national level for individual classes. The same goes for teachers' capacity-building development.

Context of Vietnam Education

Teacher Continuing Professional Development (CPD) Service

A large number of higher education teachers in Vietnam are facing many technological challenges: the lack of proper digital education and competence, the "drastically rapid-changing nature of education, and the fluid learners' value chain" (Decision No. 749/QD-TTg, 2020) in various digital learning environments that require a new complex of teacher's competencies (Decision No. 89/QD-TTg, 2019).

Table 14.1 Statistics of Education in Vietnam (School Year 2018–2019)

Total Higher Education Institutions/ Teacher Education Institutions	257/153
Higher Education Teachers	73.312
Higher Education Students	1.526.111
Master Students	97.134
Ph.D. Students	11.000
Schools (General Education)	27.723
General Education Teachers	826.704
General Education Students	16.525.868
Total Population in Vietnam	96.208,984

Source: Ministry of Education and Training (MOET), Vietnam https://moet.gov.vn/thong-ke/Pages/to-gap.aspx.

The rest, "digital natives" (born and exposed to a digital society with easy access to technology), are more individualistic with "their learning and self-orientation, regardless of the content" (Linda Daniela, 2019).

As technologies can support, affect, and even change ways of living, they surely can change how we interact with education (Table 14.1).

According to the Education Law of Vietnam, university lecturers are educational officials responsible for 1760 working hours per year, including teaching, researching, community service, and CPD. For university lecturers, the characteristics of CPD work are self-study, self-research, and participation in professional academic activities. The National Education Program mandates K–12 teachers accomplish 10 teaching months per year and 120 hours of CPD refresher organized by the Ministry of Education and Training (MOET) / Department of Education and Training (DOET) / BOE and teacher education institutions to cover updated pedagogy.

In the context of COVID-19 (school years 2019–2020 and 2020–2021), Vietnam had 79.7% of students studying online, greater than the overall average of OECD countries (67.5%). About 80% of university institutions have organized online learning on different Learning Management System (LMS) platforms; many have implemented full-online or blended delivery. Lessons were delivered mostly via virtual learning platforms.

However, online formats have yet to be uniformly implemented due to a lack of digital resources and an underutilized LMS. The teaching staff is not sufficiently equipped for online mode in terms of methods and accessibility, especially in remote areas.

Policy for Digital Education Transformation

In early February 2020, the Government of Vietnam issued a decision on the announcement of the COVID-19 epidemic (Decision No. 173/QD-TTg, 2019): institutions in all 63 provinces and cities in Vietnam stopped face-to-face learning. This was the first time that online teaching was ever deployed nationally. MOET then established a "contemporary disrupted schools while undisrupted class" motto and directed all Departments of Education and Training (DOET) with higher education institutions to employ online and proactive education. Instead, 100% of institutions shall launch remote educational activities, in which at least 20% of the curriculum shall be taught online experimentally.

Re-shaping Education in Digital Transformation

Transforming from Hybrid to HyFlex Models in Learning

The maturity model of technology adoption in educational organizations (MMEO) to trace digital transformation in educational organizations suggested by Marc Egloffstein and Dirk Ifenthaler (2021) and the ideas of smart education (SmE), smart university (SmU), smart classroom (SmC), smart learning environments (SLE), and related topics (Vladimir L. Uskov, Jeffrey P. Bakken, Robert J. Howlett, and Lakhmi C. Jain, 2018) provided the initiatives to make blended learning phenomenon operationalized and administered in Vietnam practice. When discussing the concept of learning in the context of technology, Neil Selwyn et al. (2020) confirmed its prominence, "leading to *classrooms on platforms* rather than *platforms in classrooms*".

Blended Learning

Blended (or hybrid/mixed) learning is a combination of face-to-face and e-learning activities, either entirely, partially, or non-simultaneously. This approach fosters personalized processes, thereby facilitating the generous growth of digital technology. It fundamentally changes old educational perspectives:

- Redefinitions of teaching philosophy and personalized education.
- Redesign teaching environments in terms of time and space.
- Changes in methods, forms, and tools to carry out the teaching process.
- Changes in the role of learners and teachers.
- Changes in testing methods and assessment tools.

Massive Open Online Courses (MOOCs) have been a prominent trend in educational technology over the past decade, garnering significant attention from researchers and practitioners. Despite this, MOOCs have not revolutionized the higher education sector as extensively as was initially predicted. Eveline Wuttke, Jürgen Seifried, and Helmut Niegemann (2020) consider MOOCs a "promising option for work-related learning and professional development with different models underlying pedagogical approaches."

MOOCs (Massive Open Online Courses) and SPOCs (Small Private Open Courses) offer open-source code and an accessible and adjustable interface, thus allowing variedly diverse content and meeting almost all studying demands. Technology has generated some characteristics, as follows:

- Reorienting core values in education: the role of co-creators for new knowledge; learners' adaptivity, transformation, and entrepreneurship in the volatile labor market;
- The trend of developing open educational programs (interdisciplinary, transdisciplinary, closely linked with social and practical contexts);
- Increase the ability to connect infrastructure, solutions, and tools in all fields and stages of the educational process on technology platforms;
- Requirements on digital education and digital competence for teachers and learners, "digital citizens/natives," and related parties;
- A drastic change in the nature, methods (models), and outputs: comprehensive connectivity, multidimensional interaction, and approach to the "learner value chain";
- Maximum expansion and flexibility in a digital-based learning environment (Figure 14.1).

ePedagogy

ePedagogy (Digital Pedagogy) Concept

In Vietnam, teaching processes are to be designed and operated to increase strong learner orientation and "re-configuration" toward:

- a suitable combination between formal curriculum and self-oriented learning;
- internal organization and cross-setting group activities;
- Spontaneous teaching (at any time requested by personal demands).

Educational applications can help connect to large databases and multi-format digital resources (games, simulations, interactive 3D, VR/AR/MR/XR, interactive e-books, 360 interactive videos). BYOD in classrooms is a popular global trend with smart devices that allow easy sharing and interaction and many informal spaces for learning as well as classrooms. The smart classroom included approaches, features, and smartness levels addressed (Vladimir L. Uskov, Jeffrey P. Bakken, Robert J. Howlett, and Lakhmi C. Jain, 2018). However, in the context of a fully online learning mode, this concept could be transformed into BYOCH (Bring Your

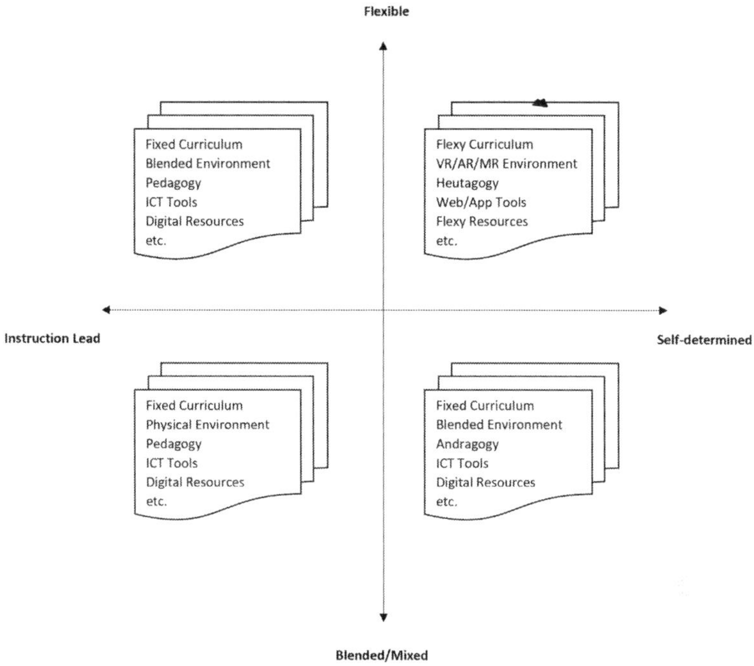

Figure 14.1 HyFlex model of learning (Hybrid, Blended / Mixed and Flexible).

Class to Home) with flexible content, activities, and devices for learners (Charles Leadbeater, Annika Wong, 2010).

The Move from Pedagogy, Andragogy, and Heutagogy

VUCAD (*Volatility, Uncertainty, Complexity, Ambiguity, and Disruption)* society promotes a new approach to CPD implementation for training staff: free room for self-identifying demands, allowing more accurately curated professional routes. Also, the *heutagogy* approach is highly relevant and likely to be realized in the digital transformation of education, especially in the form of online learning at present (Table 14.2).

Double-loop learning (dynamic contexts to adapt goals) and *deuteron-learning* (achievements, experiences, and conditions reflected regarding promoting and hindering aspects) together form an organizational structure for the CPD development model.

eLearner

According to Deloitte's analytical report (2018), the general trend shows that digital learners today are "very busy, curious, and eager to learn" (https://bit.

Table 14.2 The Move from Pedagogy, Andragogy to Heutagogy

Pedagogy, Andragogy Concept	*Heutagogy Concept*
Behaviorism	Technology-enabled Behaviorism
Constructivism	Technology-enabled Constructivism
Connectivism	Technology-enabled Connectivism
Single-loop learning	Double-loop learning/ deutero-learning
Instructor-learner directed	Learner-directed (self-determined)
Getting to learn	Getting learning self-determined
The linear design of learning	Non-linear (grid) design of learning
Competency development	Capacity building and development
Problem-solving focused	Personal development focused
Cognition/Meta-cognition	Epistemic cognition
Learner dependence/independence	Learner autonomy (fully)

ly/2KzqxFz). These overarching features can be identified in detail in the following manifestations:

- Constantly being overwhelmed and distracted by innovations;
- Tendency to be interested in learning and absorbing new things;
- Access to quality, timely responses, and personalized content;
- Considering learning as a necessary demand of all time, at all places;
- Using smartphones or mobile devices inseparably;
- Integrating relevant and future-useful learning and practical experience;
- Emphasizing the quality, easy, and suitable technology in learning;
- Being eager to learn and improve soft skills;
- Self-motivated learning for better performance and higher productivity.

With accessibility to new technology, they are increasingly becoming "the center of their learning," allowing wider parameters of self-orienting learning and turning them into "co-creators" of knowledge for "mass intelligence."

e-learning or mobile (ubiquitous) learning

E-learning provides the organization for open learning spaces and strong interoperability between participants and knowledge.

- M-learning (mobile learning): flexible teaching to maximize personal needs.
- U-learning (Ubiquitous learning): in-time responses teaching at any time, in any space, with any learning needs of learners.
- (MOOCs) and (SPOCs): accessible online learning platforms for various demands.

Due to COVID-19, most institutions in Vietnam are stopping school and applying distant learning through M-learning, proven effective in continuing education from any location, with any mobile device, at any pace: Zoom, MS TEAM, Padlet, Google Meet, LMS Moodle, Google Classroom, Classdojo, Edmodo, or YouTube videos.

Personalized Learning in CPD

Charles Leadbeater and Annika Wong (2010) stated that even in parts of the developed world, sustained investment in schools and teachers has not led to expected improvements in educational outcomes. By analyzing the matrix of school renovation (the characteristics of innovation aspects and learning modes), they argued that school improvement on its own will not be enough to meet the need for learning.

Personalized teaching is gradually becoming an important "form" of the social system in teaching, as all participants contribute to creating knowledge where they focus on the significance of informal learning to develop a new logic of learning (i.e., learning as production, problem solving, learning from peers, and/or without teachers).

In essence, this new relationship reshapes teaching by providing flexible learning for individuals' needs. It should not be one-size-fits-all. Still, this concept is open to debate, specifically:

- Must come from the capacity, the actual level, and the ability to decide for the actions of the learners (learner agency);
- Diversify teaching methods and methods to create timely intervention support when learners need it (supportive, highly adaptive teaching);
- Contextualize teaching in a multidimensional relationship and authentic assessment (based on the evidence of learners' performance).

Technology-enabled Personalized Learning

The smart education model (SMART Education) enables personalization, which sets mutual effects, promoting quality and self-efficiency, including S (self-directed), M (motivated), A (adaptive), R (resources), extended learning materials, and T (technology).

- Changing learners' expectations and learners' ability to respond to evolving social contexts (self-direction, lifelong learning, etc.);
- Orienting to diversify outputs, "educational/teaching products", according to individual learning styles and learning capacities;
- Adjusting the relationship, role, and position of teachers and learners in the teaching process in the educational ecosystem;
- Expanding space, teaching environment, and learning campus to meet the diverse needs of learners;
- Changing management and administration models in teaching based on individual learners's learning plans and pace;
- Designing teaching programs of flexibility and adaptability (learning content and assessment), focusing on learner objectivity and capacity development.

Based on the SMART model, it is possible to deploy non-traditional teaching methods and comprehensively focus on individual learners. However, Linda Daniela (2019) also warned about the risks of using technology that is not appropriate for pedagogical contexts or principles. This could lead to the so-called "*centrifugal effect*" which may call for smart pedagogy.

Challenges of High-Performing Digital Education Ecosystem

Establishing Digital Systems and Policies

The first challenge of a high-functioning digital education ecosystem is how to establish a functioning digital system and policies to carry it out. Valerie Hannon, Alec Patton, and Julie Temperley (2011) suggest that fostering a thriving ecosystem of innovation should be the goal of education. This approach will help organized learning adapt effectively to the pressures and opportunities for change. The use of the heuristic introduced above (*heutagogy*) can give an indication of the overall health of the digital innovation ecosystem.

The system: Information Communication Technology (ICT) implementation should be thoroughly integrated into all aspects of education processes, with standards precisely set up and forecasted for debugging.

The policies: Evidence-based guidelines for "digital education interventions" must be crafted with best practices and plans to develop teachers' ICT and digital literacy.

The trends: education processes (i.e., concepts, approaches, methods, etc.) emerge over time and broaden the learning and teaching dimensions.

Digital popular educational trends (virtual, augmented, and mixed reality);
Big data, learning analytics, AI, and intelligent support;

Personalized or individualized learning adapted to the contextualized, situated needs of individuals and groups;
MOOCs, SPOCs, virtual learning environments (VLEs), and LMS for larger audiences of lifelong learning programs (Eveline Wuttke, Jürgen Seifried, Helmut Niegemann eds. 2020).

Teachers are the baseline factor for systematically transforming the digital system and transcending its limitations. Thus, CPD models need to be designed as mini-credential modules that adhere to Teacher Professional Standards for recruitment and appointment in different localities and are regularly updated every year.

Enhancing Competencies and Skills Development and Optimizing Teacher Roles in the New Context of Digital Transformation

As *teachers* are the key factor in making this transformation applicable, institutions need to realize that investment in human resources and time is necessary to properly utilize digitized systems for improving learning process quality and effectiveness.

The development of teachers' competencies is crucial to supporting the shift of fixed-traditional schools' systems format toward integrated online, more flexible education supports for lifelong learning, non-formal, and informal learning.

Preparing Future Teachers for Digital Education

The goal is to ensure that teachers are ready for an agile situation and further develop digital education systems.

The OECD (2019), by recommending the vectors of the digital transformation, explored that it is not a simple phenomenon but a complex range of continually unfolding, interrelated, and often unpredictable developments. Furthermore, the transformation also challenges traditional thinking on how economic and social activities are most effectively organized and performed within the digital education ecosystem. It is not simply a question of whether public institutions, private enterprises, or non-profit organizations are best suited to a particular task in teacher preparation.

The government's policy has particularly emphasized the priority of enhancing and developing the professional expertise and capabilities of the faculty in the coming years, with the aim of creating a smart education ecosystem for Vietnam's education (Decision No. 89/QD-TTg, 2019). They need to be given the role of "transforming teachers, managers, and social actors" in the context and the mastery of incorporating technologies into educational processes to meet the demands of learning at the time.

Accordingly, teacher capacity building for the time being must develop and adopt pre-service and in-service training policies, professional accreditation, and standards on digital literacy and application. This helps education reach out more widely to rural and remote communities, breaking the barrier of basic infrastructure through solidary economies, convivial technology, and respectful design (Neil Selwyn et al., 2020).

Institutional Challenges

In the transformation process, the current status of academic institutions also posed challenges to the systematic implementation:

- Ownership (or possession) of standard technological infrastructure for implementing IT systems;
- Framework for organizational collaboration, performance management, and strategic training programs to correspond to these powerful resources;
- Appropriate curriculum and course contents of the currency and up-to-date methods, including flexible, student-centred approaches…;
- Pre-service and in-service training, an accrediting system, and professional development courses (micro credential) to innovate in the digital teaching and learning environment;
- Deployment of the appropriate technology, in particular, software and interface design, authoring tools, digital lessons and multimedia content, smart devices, etc.;
- Managing and ensuring the stability, reliability, availability, and affordability of the required technological innovations over time;
- Other than the technical challenges, institutions also need to address strategic development concerning three main components: administrators, teachers, and students; the complex interaction between these three factors; and understanding the previous technological use to help all adapt and be ready for the new technologies.

Conclusion

Smart education and the educational ecosystem are becoming an inevitable trend in the general education context of the 21st century. The digital transformation in education is restructuring to innovate for future education, providing an open, lifelong learning opportunity for everyone. The understanding of high-performing digital education ecosystems proposes that leaders integrate solutions to truly transform and sustain a system-wide education organization.

Digital education is a trend that will continue to thrive in the following decades. The rapid development of digital emergent and convergent technologies, artificial intelligence, big data, and blockchain is fostering changes in learning and providing growth opportunities in education: adaptive learning, analytics support, self-determined learning, and autonomy for the future.

The increasing use of digital educational technologies in the learning process is an upcoming, evolving, and currently underserved area for teacher training, capacity building, and development.

References

Ben Williamson, Rebecca Eynon, John Potter (2020). Pandemic politics, pedagogies and practices: Digital technologies and distance education during the coronavirus emergency, *Learning, Media and Technology*, 45:2, 107–114, https://doi.org/10.1080/17439884.2020.1761641

Charles Leadbeater, Annika Wong (2010). *Learning from the Extremes*, Cisco Systems, Inc. Cisco Public Information.

Decision No. 89/QD-TTg (2019) approving the Proposal for enhancing the competencies of lecturers and administrators of Higher education institutions meeting requirements for radical changes in education and training from 2019 to 2030.

Decision No.173.QD-TTg (2019) declared the acute respiratory disease caused by the new coronavirus (nCoV) an epidemic in Vietnam.

Decision No.749/QD-TTg (2020) approving the National Digital Transformation Programme by 2025, with orientations toward 2030.Dirk Ifenthaler, Sandra Hofhues, Marc Egloffstein, Christian Helbig (2021). *Digital Transformation of Learning Organizations*, Springer Nature Switzerland AG, https://doi.org/10.1007/978-3-030-55878-9

Eveline Wuttke, Jürgen Seifried, Helmut Niegemann (eds.) (2020). Vocational Education and Training in the Age of Digitization. Challenges and Opportunities, Förderung durch den Open-Access-Publikationsfonds der Goethe-Universität Frankfurt, http://dx.doi.org/10.3224/84742432

Linda Daniela (2019). *Didactics of Smart Pedagogy for Technology Enhanced Learning*, Springer Nature Switzerland AG, https://doi.org/10.1007/978-3-030-01551-0

Neil Selwyn, Thomas Hillman, Rebecca Eynon, Giselle Ferreira, Jeremy Knox, Felicitas Macgilchrist & Juana M. Sancho-Gil (2020). What's next for Ed-Tech? Critical hopes and concerns for the 2020s, *Learning, Media and Technology*, 45:1, 1–6, https://doi.org/10.1080/17439884.2020.1694945

OECD (2019). Vector of digital transformation. OECD Digital Economy Papers, No. 273.

Roy Schwartzman (2020). Performing pandemic pedagogy, *Communication Education*, 69:4, 502–517, https://doi.org/10.1080/03634523.2020.1804602

Shahnawaz Saikat, Jaspaljeet Singh Dhillon, Wan Fatimah Wan Ahmad, and Robiatul A'dawiah Jamaluddin (2021) Systematic review of the benefits and challenges of mobile learning during the COVID-19 pandemic. *Education Sciences*, 11, 459. https://doi.org/10.3390/educsci11090459

Vladimir L. Uskov, Jeffrey P. Bakken, Robert J. Howlett, Lakhmi C. Jain (2018). *Smart Universities Concepts, Systems, and Technologies*. Vol 70. Springer International Publishing AG.

Valerie Hannon, Alec Patton, Julie Temperley (2011). *Developing an Innovation Ecosystem for Education*, Cisco Systems, Inc. Cisco Public Information.

Chapter 15

Acceleration of Active Learning in Teaching Learning Transactions

G. Singaravelu

Objectives

1. To accelerate teaching and learning through active learning
2. To learn the concept of active learning
3. To implement active learning
4. To ensure the need for active learning strategies
5. To understand activities techniques in the classroom transactions
6. To understand the active learning method
7. To learn active learning in technology
8. To perceive the challenges of active learning in classroom learning

Introduction

Active learning happens with the participation of students in an active manner and interacts with the teaching-learning process as opposed to passive learning. It gives more opportunities to engage actively with the information when students are learning and perform better than passive learning. Active learning is the need of the hour to introduce among the students in the pandemic situation, and it encourages to activate the brain, which accelerates cognitive and sensory networks, which helps

 DOI: 10.1201/9781003504894-15

the process of new information and incorporates a wide assortment of homeroom exercises which is presented before the students who enable insight and understanding the concept clearly. Active learning is a dynamic learning approach that assists with expanding knowledge, motivation, understanding, and retention. Another approach of AL is a flipped classroom. The entire thought of a flipped homeroom depends on dynamic learning. In a flipped classroom, video addresses and readings present the fundamental realities. Few related studies support the essential realities of genuine comprehension with the direction of educators who support AL for effective learning. There is a grounded proof base supporting the utilization of active learning to enable the students to learn more with retention. The advantages to utilizing such exercises are many, including further developed basic reasoning abilities, expanded maintenance and movement of new data, expanded inspiration, worked on relational abilities, and diminished course disappointment (Prince, 2004).

Meaning of AL

Active learning activates the students in a wide scope of educational techniques, which draw in understudies as dynamic members in getting the hang of during the class time with educators. Commonly, these techniques speed up the understudies' cooperation during class; however, they may likewise include individual work as well as reflection. AL approaches ranged from short, straightforward exercises like diary composing, critical thinking, and matched conversations, to longer, involved exercises or academic systems like contextual investigations, pretend, and organized group-based learning. AL depends on conversation, practice, survey, or application. Problem-settling capacity and investigating new ideas in bunches AL requires understudies to take part in significant exercises and ponder the ideas on learning. The maintenance force of the understudies is expanded more in active learning by cooperating to apply other training. AL is regularly differentiated from the customary talk where understudies latently get data from the educator.

Aims of AL

- Making students learn effectively the concept in a simple manner.
- Providing freedoms to foster an efficient arrangement and basic consciousness of instructive approach, hypothesis, and practice
- Facilitating the students to perform well through interaction mode with teachers.
- Accelerating students to be active in learning continuously for a long duration.

- Escalating understudies' learning through contextual analyses, pretends, and group-based learning.
- Affording the information and abilities that empower understudies to apply hypothesis and practice.
- Enabling understudies to accomplish a fitting degree of subject information for application in foundations.
- Developing comprehension of the standards and practice of evaluation identified with the National Curriculum Framework and details of assessment. Encouraging students to learn simple from complex.
- Ensuring the retention power of the students.
- Achieving the concept of learning by doing.
- Increasing ability to resolve complex issues by interaction mode.
- Understanding the different Strategies in active learning
- Practicing the active learning strategies, methods, and approaches.
- Creating an appropriate learning environment for the students.
- Implementing active learning in all levels of institutions.

Principles of Active learning

Barnes (1989) suggested the following AL principles

1. **Purpose:** significant assignment.
2. **Reflection**: understudies' appearance on the importance of what is realized.
3. **Negotiation:** exchange of objectives and strategies for learning among understudies and educators.
4. **Critical:** understudies like various available resources for learning the substance.
5. **Complex**: understudies contrast learning undertakings and intricacies existing, all things considered, and making intelligent investigations.
6. **Driven-Situation:** the need of the circumstance is considered to set up learning assignments.
7. **Engagement:** genuine errands are reflected in the exercises directed for learning.

Benefits of Active Learning

The advantages of active learning are various because it draws in understudies effectively. In a traditional addressing technique, it is simple for understudies to do the "proper thing" (keep their eyes on the instructor, take notes, and so forth), all while wandering off in fantasy land and not contemplating the material. In dynamic learning, understudies are in more control of the learning and, as the very name proposes, dynamic and drawn in with the material. It is valuable since it moves the accentuation on gaining from the instructor to the understudy. Understudies are continually communicating with the material and making a unique interaction with the substance. In doing this, understudies rapidly move from straightforward transient maintenance to using higher-request thinking abilities which assists understudies with fostering a more grounded comprehension of the material and applying it in certifiable situations. The advantages of dynamic learning have been upheld consistently in the writing. By contrasting understudy learning gains in early physical science courses, Richard Hake had the option to show that intelligent courses were more than two times as powerful in advancing applied arrangement when contrasted with conventional ones (Hake, 1998). A later meta-investigation by Freeman announced outcomes from 225 examinations across STEM disciplines, contrasting conventional talk with dynamic learning (Freeman et al., 2014). As a rule, understudies' normal test scores were displayed to improve by around 6% in dynamic learning classes. Furthermore, understudies engaged with customary talk were observed to be 1.5 occasions bound to flop when contrasted with those in classes with huge dynamic learning. It requires some investment and imagination to viably fuse dynamic learning systems into educating and accomplish the full advantages across educational settings and disciplines. However, as a large number of the employees we profile illustrate, dynamic learning can undoubtedly and viably be fused into existing courses and materials without the requirement for an emotional upgrading of the course.

Activities of Active Learning

Michael Prince clarifies, "Dynamic learning is for the most part characterized as any educational technique that draws in understudies in the learning system. To put it plainly, dynamic learning expects understudies to do significant learning exercises and contemplate what they are doing" (Prince, 2004). Sovereign's definition is drawn from Bonwell and Eison (1991) and has been broadly acknowledged. Dynamic learning exercises assist with advancing higher-request thinking abilities like the use of information, examination and combination. It likewise draws in students in profound instead of surface figuring out how to move their insight in a better manner.

Active Learning Activities

- Promoting Higher Order Thinking
- Promoting the Application of Knowledge

Figure 15.1 Active learning practices.

- Promoting Analysis and Synthesis
- Engaging Deep Learning
- Enabling Application of Knowledge
- Enabling to application of Knowledge
- Enabling to Transfer of Knowledge (Figure 15.1)

AL Strategies

Active learning strategies are based on student-centered teaching. Students learn in their faces by supporting teachers. In active learning strategies, teachers are acting as facilitators, not as managers. Active learning strategies are starting to learn from simple to complex. Some effective strategies of active learning are performed in the classroom transaction. Active learning strategies are started simply through large group discussions, group evaluation, case studies, Interactive lectures, role play, Jigsaw discussion and site visits. Some of the complex strategies are self-assessment, Think-pair share, peer review, Brainstorming, Hands-on Technology, Simulation, and Inquiry learning.

Think-pair-share is an incredible strategy that pairs with an accomplice to talk about, and just offer with the entire gathering. Another dynamic learning system is a sensation that can be utilized in the homeroom. Students of Kinesthetic are those understudies who learn by moving their bodies and contacting manipulatives.

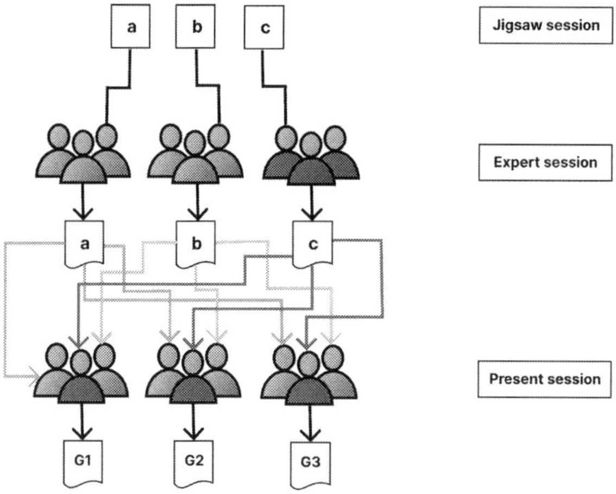

Figure 15.2 Active learning strategies.

Generally speaking, utilizing dynamic learning procedures is one of the best approaches to establishing a difficult learning climate where all understudies are roused to make a valiant effort to succeed (Figure 15.2).

Exercises in AL Strategies

The following exercises are active learning strategies:

- Debating a recent development
- Engaging if reviews or reproductions
- Actively self-checking their learning
- Constructing a test for use with peers
- Reflecting on a point shrouded in class
- Interviewing specialists in their field
- Learning through Play
- Role Play
- Debates
- Group Projects
- Peer Teaching
- Think-Pair-Share
- A Kinesthetic Approach
- Grab Bags
- Phenomenon Based Learning
- Inquiry-Based Learning
- Discovery Learning

- Challenge Based Learning
- Game playing technique
- Game-Based Learning
- Guided Practice

Reasons to Use AL Strategies

Main reasons to use AL strategies:

- AL in the study provides guarantees that every understudy has a voice in their learning, subsequently expanding understudy commitment throughout the functioning day.
- Illustrations of dynamic gaining from an assortment of study halls show that understudies who are presented with dynamic learning systems feel like they know themselves better as students and masterminds. Such understudies are better prepared to expand their own adapting autonomously throughout the process of everything working out.
- Students can be active throughout the hour in the classroom in adopting AL-based learning (Figure 15.3).

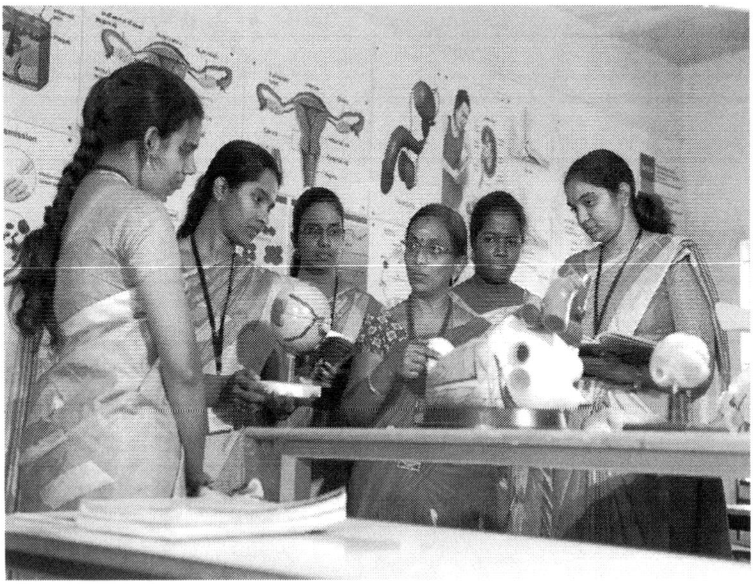

Figure 15.3 Using active learning.

Table 15.1 Active Learning Techniques

Cooperative Learning Techniques – Base groups, visual diagrams, thing/pair/ share, write/ share, Jigsaw, affinity diagrams, team discussion, vision exercise, etc.	**Technology** – Helps students manage all the different aspects of the activity • Word Processing • Presentation tools • Spreadsheets • Social networking tools • Chat rooms
Collaborative Learning – Consensus groups, peer writing, peer tutoring, etc.	• Cases • Mini Lectures

Active Learning Techniques

- The think-pair-share movement is a functioning learning strategy which urges all understudies to interface with the material. In this movement, the educator poses an open-finished inquiry.
- Using a Minute Paper in the class is a valuable developmental evaluation.
- Peer Instruction is likewise a strategy, educators stop during class and ask understudies a calculated inquiry.
- Asking understudies to cooperate in bunches is an extremely compelling approach to effectively draw in them with coursework.
- A group of understudies can examine contextual analyses to apply course content to tackle genuine world issues (Table 15.1).

Active Learning Method

Active Learning Methodology and Activity Based Learning are two fascinating showing strategies which are very not quite the same as conventional techniques for instructing. The fundamental appealing marks of these strategies are that they are alluring and successful. Idea maps, charts, outlines, networks, maps, also as pictures, can be utilized as visual instances of conceptual ideas.

Idea maps, specifically, are worried about connections among thoughts. An idea map shows associations that exist between terms or ideas canvassed in course material. Understudies work in gatherings to develop idea maps by interfacing individual terms by lines which show the connection between each set of associated terms. The greater part of the terms in an idea map has different associations. Fostering an idea map requires the understudies to recognize and sort out data and to build up significant connections between the snippets of data.

Synergistic composing a community-oriented composing task can be coordinated in more than one way, particularly by trying not to give one individual everything. A gathering might be approached to separate a composing task into parts and every understudy will compose their part and carry it to the gathering for aggregation and altering.

Conceptualizing is an interaction for creating clever fixes to issues. Conceptualizing works by zeroing in on an issue, and afterward intentionally concocting whatever number arrangements would be prudent and by pushing the thoughts quite far. One reason it is so compelling is that the barnstormers not only concoct ground-breaking thoughts in a meeting but also start relationships with others' thoughts by creating and refining them. Understudies can utilize the meeting as a chance to make associations, unreservedly partner, and perceive that they have been drawing in with the subject in manners they might not have known. Ongoing examination recommends that people in little gatherings learn better compared to those who do all alone or in seclusion. Considering the examination on dynamic learning, this isn't unexpected, as a little gathering starts synergistic learning and its subsequent exercises: understudies create questions, talk about and come to results, transform thought into composed or oral language, and so on.

Active Learning in Technology

1. **Collaborative Virtual Classrooms**

   Communitarian virtual homerooms make internet learning seriously captivating. Besides the standard sound video conferencing and talk highlights, virtual study halls likewise give coordinated and offbeat explanations, correspondence, and asset sharing for facilitators and members.
2. **Mind planning/Brainstorming**

   In active learning strategy, two methodologies are grouped, in which Mind planning and conceptualizing are staple philosophies for any critical thinking action. In these classes, students concoct thoughts and post them on a board. Collectively, the understudies select the best ones and go through those to accompany an answer. For these strategies, there are accessible applications that permit students to utilize their gadgets and work together with others in thinking of a brain guide or thought tree.
3. **Scavenger Hunts**

   Scavenger Hunt is a fun and drawing action that includes the utilization of the organization's information base. The undertaking of the student is to utilize the framework and track down the right asset to discover the issue. Not only does it acquaint the students with the framework, but it also makes the understudies take on genuine situations of students.
4. **Role Playing**

   Pretending is likewise one more successful methodology established on dynamic learning methods. It recreates genuine circumstance that requires critical thinking abilities. It has a uniqueness to measure real execution.

Pretending exercises can incorporate reproductions with cooperation through telephones, email, talk, or sometimes, augmented reality.

5. **Data and Tools for Problem-tackling**

   Exercise in forager chase and pretending action are the more successful dynamic learning systems for students. The facilitator allows a contextual investigation of a student. The student, thus, sorts out the information and utilizes the accessible assets to address the case.

6. **Online Discussion Boards**

   Online conversation sheets are validated compelling dynamic support procedures. Online sheets are virtual sheets where understudies can adapt cooperatively. Students post inquiries and answer questions. More often than not, there is next to no facilitator or well-informed authority mediation required, with most answers generally coming from different members who are more educated on the theme. A case is appointed by the facilitator by the student.

7. **Learning by Teaching**

   Learning by instructing alludes to instructors permitting students to get ready and show examples to their kindred understudies. Even though it might appear as though the facilitator is adopting an extremely uninvolved strategy in this technique, it includes an exceptionally intricate interaction where the facilitator is both a mediator and an educated authority.

8. **The Jigsaw Technique**

   The Jigsaw Technique is one more way to deal with receiving the rewards of dynamic investment and cooperative learning. In this methodology, students are given a "piece of the riddle" that they need to tackle all alone. After this, they need to team up with different students to at last finish the puzzle. This approach would be a decent expansion to pretending and utilizing information/apparatuses to take care of more serious issues as well as giving members a brief look at the "master plan." It is a decent exercise to allow students to understand their job in the master plan by doing both individual and community work and how those are all important for an interaction (Figure 15.4).

9. **The 'Flipped Classroom'**

   The flipped homeroom is a genuinely new term in the learning and Ed-tech industry. Illustrations are 'flipped;' implying that the greater part of the work like perusing and exploration are completely done outside of class. It goes as opposed to the customary methodology where the majority of the class time is utilized for talks, and exercises are doled out as schoolwork. Flipping a study hall allows for the facilitator to execute dynamic learning strategies during class time.

10. **Game-Based Learning**

    Game-based learning is, apparently, the best time among all the dynamic learning techniques. Game-based learning, or gamification, is transforming a specific part of learning into a game.

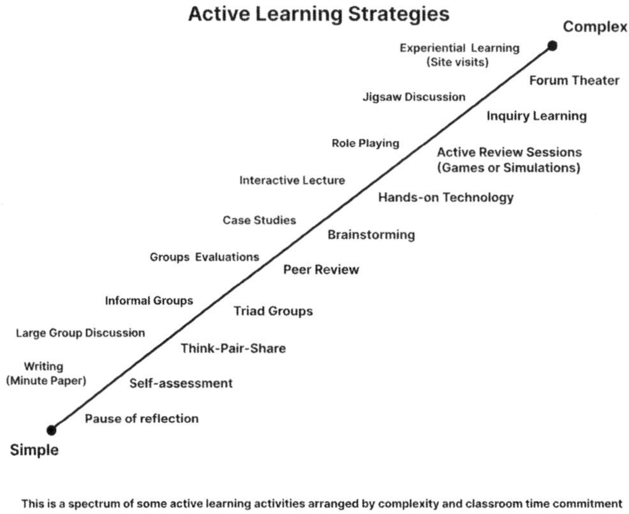

Figure 15.4 Groups in Jigsaw technique.

Table 15.2 Active Learning and Pedagogy

Bloom's Original Taxonomy		*Revised Bloom's Taxonomy*
Evaluation		Creating
Synthesis		Evaluating
Analysis		Analyzing
Application		Applying
Comprehension		Understanding
Knowledge		Remembering

Active Learning in the Classroom

In the dynamic learning study hall, understudies are occupied with the learning system and wrestling with the course material by working cooperatively on issues. Understudies are continually handling learning and checking out PowerPoint slides, yet they are likewise examining or discussing questions presented by the teacher and are asked to investigate the data introduced. It is an understudy-driven methodology to enable, connect with, and animate a study hall by putting understudies at the focal point of the learning system (Table 15.2).

Advantages of Active Learning

- Encouraging the interest of the students.
- Accelerating motivation of the students.
- Increasing students' involvement.
- Allowing free expression of ideas and opinions.
- Permitting practice skills of students.
- Improving dynamic group activity
- Recognizing a variety of learning styles.
- Encouraging students to take responsibility.
- Ensuring self-learning.

Active Learning Challenges

The main challenge in AL is delimiting the speed of the homework. All the dynamic exercises for learning can be separated effectively and students are activated to learn by using material at their speed. Further perfect measurement of the program of the study should not divert the attention of the students. Active learning works will be more achieved than other methods of learning. Clear destinations assist instructors with developing and working with exercises that permit substitutes to accomplish dominance while additionally figuring out how to take an interest in gatherings and their self-face learning. Though there are challenges with active learning, they can certainly be overcome with thoughtful planning. Some challenges of active learning are No focal point, multiple distractions, overwhelming Teaching, Adjusting the challenges of AL classrooms, encouraging reluctant students, and developing a communal philosophy for improving active learning.

Conclusion

Perceiving the meaning of active learning, method of AL, and activities of AL are need of the hour to students and teachers for making effective classroom with conducive atmosphere. Implementing AL in the classroom transaction eliminates the difficulty of the learners in difficult areas of the various subjects. Sensitizing the challenges of AL is mandatory for the teachers to find out the appropriate pedagogical activities for supporting learners to enable the students to acquire knowledge in an effective way.

References

Barnes, D. R. (1989). *Active Learning.* United Kingdom: Leeds University TVEI Support Project.

Bonwell, C. C. & Eisen, J. A. (1991). *Active Learning: Creating Excitement in the Classroom*. School of Education and Human Development, George Washington University: Washington DC.

Freeman, S., Eddy, S. L., McDonough, M., Smith, M. K., Okoroafor, N., Jordt, H., &Wenderoth, M. P. (2014). Active learning increases student performance in science, engineering, and mathematics. *PNAS Proceedings of the National Academy of Sciences of the United States of America*, 111(23), 8410–8415. https://doi.org/10.1073/pnas.1319030111

Hake, R. R. (1998). Interactive engagement versus traditional methods: A six-thousand-student survey of mechanics test data for introductory physics courses. *American Journal of Physics*, 66(1), 64–74.

Prince, M. (July 2004). Does active learning work? A review of the research. *Journal of Engineering Education*, 93(3), 223–231. https://doi.org/10.1002/j.2168-9830.2004.tb00809.https://www.shutterstock.com/image-vector/brainstorm-concept-mind-map-creative-think-436670707

Chapter 16

Revitalizing Education

Building Educational Resilience in the Post-Pandemic Era

A. Sivakumar and V. Vanitha

Objectives

1. To analyze the global impact of the COVID-19 pandemic on education, considering the extent of disruption and its consequences for various student populations.
2. To examine the role of technology in reshaping the educational ecosystem, particularly in the context of the pandemic, and assess its effectiveness in facilitating remote and online education.
3. To explore the challenges and opportunities faced by educators and students during the transition to remote learning, with a focus on issues such as student motivation, engagement, and the digital divide.
4. To investigate innovative teaching and learning approaches, including the flipped classroom model, blended learning, and hybrid instruction, and their potential to augment educational standards in the post-pandemic era.
5. To assess the need for ongoing teacher training and support, considering the evolving role of educators in technology-driven learning environments and the importance of student-centered approaches.

Introduction

The COVID-19 pandemic crisis led to a substantial response, with over 100 countries enacting legislation to close schools in March 2020. This sweeping measure

DOI: 10.1201/9781003504894-16

had a profound impact, fundamentally changing the educational experiences of more than 860 million children and youth worldwide (Viner et al., 2020). The global COVID-19 pandemic had profound and unsettling educational repercussions on systems worldwide, presenting both challenges and opportunities, particularly for vulnerable populations like economically disadvantaged students, marginalized individuals, and those individuals with diverse abilities. This impact emanated not only from the immediate medical ramifications of the pandemic but also from a cascade of subsidiary consequences, including reduced family incomes, nutritional instability, heightened abusive relationships, and various other regional and social repercussions. The disruptive effects of the pandemic were staggering in scale, impacting over 1.7 billion learners globally. Among these learners, a staggering 99% were students from economically vulnerable countries, underscoring the widespread reach of educational disruptions (The Organization for Economic Cooperation and Development (OECD, 2020c; United Nations, 2020, p.2).

The coronavirus pandemic has brought about a paradigm shift in universal education through the groundbreaking introduction of innovative resources, aids, and approaches. According to UNESCO's data, the global cessation of university operations in response to the COVID-19 pandemic has impacted nearly nine out of every ten students. The sudden onset of the pandemic and the subsequent closure of schools prompted a massive and rapid response from educators and education systems worldwide. These adaptations and innovations were necessitated by the urgent circumstances at hand. In a remarkably short period, many educational institutions and systems swiftly transitioned to offering remote education (Kamanetz, 2020; Sun et al., 2020). This shift in educational delivery methods took various forms, including television and radio broadcasts, online platforms, or even traditional postal services, as schools endeavored to reach and teach students in diverse ways. Regardless of the outcomes, remote learning became the default mode of education provision for varying durations. Educators demonstrated proactive responses and provided substantial support for the transformation in lesson delivery.

The remarkable advancements in the field of education have prompted the stakeholders to embrace digital strategies both domestically and on-site, as noted by S. J. DeLozier and Rhodes (2017). The utilization of digital platforms for education became the most viable option to sustain the teaching-learning process during the COVID-19 pandemic. With the need for people to isolate themselves to prevent virus infection and transmission, this virtual medium was essential to ensure that students could pursue their education without adversely affecting their academic progress. But it has also given rise to doubts, as expressed by R. Davis in 2007 (Davis & Wong, 2007), regarding whether the education system had incorporated sufficient flexibility and whether students and faculty possessed the adaptability, decision-making abilities, problem-solving skills, and, most importantly, the capacity to navigate technological advancements effectively.

However, as noted by P.-S. Goh in 2020 (Goh & Sandars, 2020), today's digitally driven generations exhibit a high degree of comfort with online platforms, potentially paving the way for E-learning to become the predominant mode of education, particularly in situations where on-campus learning is impractical. Overcoming contemporary difficulties may ultimately foster an enhanced forthcoming learning environment.

Technology: A Catalyst for Change

In the 21st century, technology has become an indispensable component of the learning process. The increased integration of technology in education has transformed educators' roles from traditional knowledge distributors to more flexible positions as supporters and motivators, encouraging active student participation and learning, as noted by E.M. Onyema et al. in 2019. Furthermore, as suggested by Shadiev and Sintawati in 2020, technology plays a crucial role in fostering intercultural learning at various levels. Additionally, technology can serve as a facilitator for personalized learning, enabling students to achieve more effective learning outcomes, as emphasized by L. Zhang et al. in 2020.

E-Learning, as a forward-looking educational approach, caters to the diverse needs and forecasts of various users. In this context, it facilitates the implementation of a range of digital technologies, reshapes pedagogical approaches, and enhances outcomes and proficiency to align with the foremost aspects of eLearning, as highlighted by H. Hamidi in 2018 (Hamidi & Chavoshi, 2018). Digital transformation is an ongoing process that signals the integration of digital technologies into our social reality. Describing the current era as "post-digital" acknowledges both the established and continually evolving nature of digitalization and its transformative influence on society, a term widely used in various studies (Jandrić et al., 2018). Artificial intelligence has a wide range of applications in education, including personalized teaching, interviews, group and individual instruction, library services, data management, and more. Its potential in education is continually expanding, though its impact, both positive and negative, remains a subject of debate.

Intelligent Education System (IES) is a common AI-driven method in education, utilizing step-by-step teaching approaches tailored to individual needs and learning progress, divided into domain models, pedagogy models, and learner models (Holmes et al., 2019). Skulmowski and Rey (2020) discuss COVID-19 as a catalyst for the digital transformation of higher education and anticipate significant reform efforts arising from the increased exposure to technology during the pandemic. Laufer et al. (2021) urge educational leaders to transition from emergency online learning to long-term, inclusive digital education visions that emphasize collaboration over individual gains.

Lessons from the Pandemic and India's Vision

The COVID-19 pandemic caused significant education disruptions globally, exposing issues like the digital divide and low digital literacy. The Indian National Education 2020 Policy Brief stressed that technology alone is insufficient for effective learning; it requires relevant content, effective teaching practices, and supportive environments. Teacher development and professionalism are essential for quality education. The post-pandemic landscape revolves around the rejuvenation of education, human interactions, and overall well-being. The crisis served as a stark reminder of the fundamental importance of public education in societies, communities, and individual lives. Goal 4 of the Sustainable Development Goals (SDGs) underscores the significance of education as it equips people with the knowledge and skills necessary for health, employment, and promoting tolerance. It strives to guarantee access to quality education for all.

Discussing the significance of enhancing the ability to incorporate the Sustainable Development Goals (SDGs) into planning and monitoring, particularly in the context of the pandemic, the NITI Aayog emphasized: During the COVID-19 pandemic, it became evident that there is a crucial need to enhance our capacity to comprehend how different aspects of development are interconnected. Furthermore, it is essential to grasp the extent to which adverse events in one sector can impede and even undo progress in nearly all aspects of the SDGs.

The NEP 2020 underscores the critical role of education, stating that it is fundamental for unlocking human potential, fostering equity, driving national development, and positioning India as a global leader. It emphasizes that universal access to high-quality education is essential for India's progress in economic growth, social justice, scientific advancement, national unity, and cultural preservation. The NEP views universal high-quality education as the pathway to harnessing the nation's talents and resources for the benefit of individuals, society, the country, and the world.

Sam Pitroda emphasizes that education is transformative and redefining, leading to enlightenment and lifelong prosperity. This underscores the importance of preparing for the future. The pandemic demonstrated the increased role of technology in education, ushering in a new digital paradigm for learning. This has given rise to a hybrid educational model where instruction is delivered through both physical and virtual formats.

The transition from traditional teaching to digital education requires a multifaceted approach. Initiatives like the National Educational Technology Forum (NETF) under NEP 2020 and UGC's push for digitization during the pandemic have been instrumental. These efforts align to reinvent the educational system through disruptive technology, as outlined in Digital India Mission and NEP 2020. Rajesh Panda's idea of a model policy on Education Technology in India further supports the growth of the education ecosystem for more effective learning.

The 2019 Annual Status of Higher Education in States and Union Territories report underscores these points and underscores the necessity to revamp Higher academic pursuits in India to address the demands of the Industrial Revolution 4.0, align with students' aspirations, cater to workforce needs, enhance artistry, and promote lifetime learning (Deloitte, 2019). The forward-thinking NEP 2019 aims to tackle existing limitations, promote digitization, prepare the futuristic workforce, and elevate the tertiary education system of India to international standards, competing with foreign universities. The primary challenges recognized by the All-India Survey on Higher Education (AISHE) include disparities in adapting courses to meet industry needs, both the volume and caliber of research, the development of a multi-skilled and high-quality faculty, finding a balance between education and job preparedness, the promotion of interdisciplinary learning, and the expansion of online education (AISHE, 2019, p.12).

India's New Education Policy aims to join ranks with countries like China and the US by increasing enrollment to 50%. Key goals include blending Humanities with STEM, offering flexible learning, ensuring institutional autonomy, and providing robust student support. Initiatives such as the Education Quality Upgradation and Inclusion Programme (EQUIP), Study Webs of Active-Learning for Young Aspiring Minds (SWAYAM) 2.0, expansion of Free and Open Source Software for Education (FOSSEE), fostering social and industry partnerships in all institutions, the Scheme for Trans-disciplinary Research for India's Developing Economy (STRIDE), and the Scheme for Transformational and Advanced Research in Sciences (STARS) exhibit forward-thinking approaches but require additional visionary enhancements.

Challenges and Unique Aspects of Online Education at a Massive Level

In India, the conduct of online classes on such a massive magnitude marked a significant first. The introduction of distance learning in response to the pandemic has introduced significant challenges to the teaching and learning processes within many schools. While the Department of Education has made efforts to implement various learning modalities, and teachers have strived to initiate innovative teaching strategies, ensuring student engagement and participation remains a persistent challenge (Palad, 2022). The absence of traditional classroom dynamics has made it difficult to motivate students and effectively monitor their progress. Furthermore, disparities have emerged among students in terms of their opportunities to participate in online classes, influenced by factors such as socioeconomic status, device availability, and internet connectivity. These disparities have led to a widening gap in students' learning experiences.

The introduction of online classes brought forth a range of challenges for both educators and students, even while the internet signifies a significant technological breakthrough that is remolding societies and universities globally (Volery & Lord, 2000). The context of digital education differs significantly from the conventional classroom environment, particularly in terms of learners' drive, gratification, and interpersonal dynamics (Muthuprasad et al., 2020).

Educators believe that virtual learning serves as a valuable alternative, ensuring that students can continue their education from the safety of their homes during pandemics. However, this perspective differs from that of professionals and executives who do not view virtual learning as a complete replacement for traditional classroom or in-person campus-based learning. The reason for this distinction lies in the unique and profound connection between a student and a teacher that develops when they share the same physical space. This interaction serves as a binding element that shapes the learning environment in a significant way. While online teaching is a viable alternative and a temporary option, it cannot replicate the depth of this in-person educational experience. The shift to remote education, including online exams, assessments, and advising, posed challenges for faculty with limited digital skills. Many institutions lacked the necessary resources and support. Faculty, regardless of age or background, had to adapt to remote teaching without proper training (Hodges et al., 2020). This was particularly tough for those unfamiliar with online teaching methods (Shulman, 1987) and lacking awareness of essential principles for effective online learning.

Amid the COVID-19 outbreak, the digital divide emerged as a significant obstacle to remote teaching and learning. Studies conducted in 2021 by the Azim Premji Foundation and Oxfam India revealed that nearly 60% of school children in India faced difficulties accessing online learning opportunities. Even among students attending urban private schools, half of the parents reported problems related to Internet signal and speed. One of the factors amplifying students' stress levels is the demanding nature of online lecture assignments, as they often involve the immediate application of newly acquired knowledge through online media. This challenge is underscored by research demonstrating that public health crises can lead to various psychological consequences for students, manifesting as anxiety, fear, and apprehension (Sharp & Theiler, 2018).

Family income stability is a significant factor contributing to the anxiety experienced by students during the COVID-19 crisis. This can be attributed to the heightened levels of both psychological and economic stress that individuals face in such circumstances (Liu, 2013). The situation underscores the considerable challenges teachers face in addressing both educational and facility-related issues within schools, leading to significant levels of burnout and work-related stress among educators. Meanwhile, students have shown a lack of motivation to complete their learning tasks, and parents have encountered difficulties in managing and encouraging their children's engagement in education at home. This emphasizes the vital importance of fostering strong, positive relationships among parents, teachers, and students as a pivotal factor in motivating learners to actively engage in their educational pursuits (Robosa et al., 2021). Indeed, researchers have identified that the allocation of technological resources, internet disruptions, and the scarcity of qualified educators pose significant obstacles to integrating technology into government schools (Salam, 2017).

Navigating the Pandemic Challenge

Worldwide education systems have historically strived to create conducive environments for students, with the overarching goal of promoting their holistic development. The concept of schools being a 'second home' has been widely embraced, where educators play a pivotal role in not only imparting academic knowledge but also instilling moral values. This structured educational system significantly influences a child's behavior, as schools often extend their roles to provide essential health services and psychological support (The World Bank, 2021). However, a shift occurred with the growing competition and the emergence of a business-oriented mindset in the field of education. Leading up to the COVID-19 pandemic, there was a perceptible shift away from prioritizing values and humanitarian concerns. Instead, students were inadvertently conditioned to prioritize competition, aiming for the top positions while potentially caring less about the well-being of others. This trend may have contributed to the emergence of a future generation that appears to be more self-centered and opportunistic in its outlook.

The initial phase of COVID-19 was marked by confusion and anxiety, as no one had encountered such a situation before. However, humans quickly devised methods to ensure the academic year continued without delay. It involved endless work and meetings to implement online education methods (Rapanta et al., 2020). Even in technologically challenged areas, educators worked hard to deliver knowledge within guidelines. Technology, once an aid, took center stage for learning and teaching during the pandemic.

Fuller (2021) examines the impact and disruption caused by technology during this period and asserts that the pandemic has accelerated the redefinition of ICT (Information and Communication Technology) not solely as a practical learning tool but also as a learning environment in itself. Insufficient attention has been given to how technology is reshaping the learning environment, from in-class to digital distance learning. Learners differ in their ability and confidence to manage their learning, including their choice of subjects. Margarita Langthaler highlights that the pandemic has revealed the downsides of rapid digitalization, leading to increased inequality and exclusion, especially along geographic, economic, social, and gender lines (Langthaler, 2021).

Amid the pandemic, a surge in online learning has brought forth professional development programs targeting teachers and others. These programs, facilitated through platforms like Google Meet and Zoom, primarily empower teachers to hone their 21st-century teaching skills. The integration of ICTs into education has brought about online programs to enhance teachers' ICT skills. This shift underscores ICT's importance in education, expands curriculum and assessment practices, and promotes more participatory education through collaborative online discussions. These initiatives simplify education during pandemics, prompting us to think about sustaining educational activities in such scenarios.

Teachers Post-Pandemic Resilience

Post-pandemic, teachers have faced significant challenges in adapting to the "new normal" while also adhering to safety protocols. This situation has placed undue stress on educators, particularly when dealing with students who have faced difficulties and fallen behind during the implementation of distance learning methods. It underscores the importance of school administrators taking proactive steps to provide teachers with training in new teaching practices, thereby equipping them to better handle classroom sessions in the post-pandemic era. Such preparation is essential to ensure a smoother transition and effective support for both teachers and students.

The success of virtual learning greatly depends on the pivotal role played by virtual tutors. They should gain insights into methods for adding a human touch to the course, engage learners effectively, and incorporate elements like community building, knowledge construction, and teaching involvement. These strategies foster academic success, social connections, empathy, tailored learning experiences, and student drive" (Martin et al., 2020). As indicated by Briones et al. (2022), the critical factors impacting students' academic advancement in school encompass student motivation, the creation of a positive learning environment, the effectiveness of teachers, and parenting approaches.

Consequently, when teachers are addressing concerns related to learning gaps identified among students following the pandemic, it is imperative to acknowledge and thoroughly communicate the significance of these factors to all stakeholders involved in the educational process. The COVID-19 crisis forced sudden changes in education, leaving many untrained educators struggling to create effective e-content and use technology to enhance student motivation, belonging, and satisfaction. Students also faced mental health concerns, increased academic anxiety, and reduced self-confidence due to the abrupt disconnection from peers and instructors (Arslan, 2021; Hehir et al., 2021).

Revolutionizing Education: Diverse Perspectives and Evolving Concepts

The COVID-19 situation has been a monumental opportunity for us to reshape our entire education system to be better prepared for future pandemics, ensuring sustainability. We must now consider a fresh and innovative education model capable of addressing emerging challenges. To achieve this, we must cultivate a forward-thinking educational vision centered around 21st-century skills. The infiltration of digital media into our education system has profoundly reshaped the long-standing traditional approach. Students now increasingly engage with electronic devices and digital content, leading to reduced social interaction with their peers. They rely heavily on these devices, often failing to express and develop their inherent abilities. Consequently, graduates emerge with limited knowledge, insufficient skills, and compromised ethical values, contributing to problems such as unemployment and exacerbated environmental crises. Therefore, we must refocus our educational efforts on nurturing 21st-century skills in learners. This shift will empower them to become effective problem solvers and leaders as we navigate the challenges of a world recovering from the pandemic's impact.

Multidisciplinary Pedagogical Approaches

It is widely acknowledged that our education system aims to empower learners to excel in the 4Cs (Critical thinking, Communication, Collaboration, and Creativity), fostering their inner abilities and equipping them to tackle challenges and thrive in the global landscape. A primary obstacle to this holistic development is the inflexible division of academic disciplines, often driven by disciplinary egocentrism. Nonetheless, real-world issues demand a multifaceted approach, necessitating diverse perspectives and skills from various fields to achieve effective solutions (Manandhar & Luitel, 2020).

In the post-pandemic world, education should prioritize integrated learning, where learners analyze problems from multiple perspectives to find diverse solutions. This approach goes beyond traditional textbooks and disciplinary methods. To address this, an integrated approach to learning has emerged, advocating for multidisciplinary, interdisciplinary, and transdisciplinary approaches (Liao, 2016). Additionally, the STEAM (Science, Technology, Engineering, Arts, and Mathematics) education framework promotes more progressive pedagogical approaches (Pant et al., 2020). STEAM education exemplifies this holistic approach, promoting the development of knowledge, skills, and competencies for sustainable problem-solving through collaborative exploration, research, analysis, innovation, and implementation. Students will gain the empowerment and skills necessary to think divergently and devise solutions to problems (Sousa & Pilecki, 2018). They will learn to creatively develop prototypes and effectively implement them. Throughout this educational journey, learners will transform into real-world problem solvers who understand the importance of cooperation in professional environments, enabling them to collaborate effectively with others.

Supporting Learner-Led Education:

Cooperative learning fosters both individual and group competencies, emphasizing interaction and accountability. It involves small groups developing interpersonal skills to enhance effectiveness, with each member shouldering a unique task, and promoting personal responsibility. This approach positively impacts Chinese students' motivation (Yang, 2014). The flipped classroom, known for its interactive teaching environment and high student satisfaction, allows teachers to focus on addressing learning challenges (Shafi, 2018). The literature highlights students' need for improved reasoning and critical thinking skills to become proficient and competent learners.

Ahmed Abdelhafez (2021) explores technological solutions for teacher training during the pandemic but overlooks learner-led education and the impending transformation. Incorporating technology into education can enhance participatory and conversation-based learning environments (Eckhaus & Davidovitch, 2019), leading to valuable outcomes and the promotion of a learner-centric approach (Castro, 2019). To optimize the learning process, various elements must be interconnected, such as the lecturer's competence, students' cognitive development, desired learning outcomes, teaching methods, and the anticipated learning environment. Consequently, it's essential to assess students' needs and preferences before integrating technology into the learning experience.

Fostering Self-Directed Learning

The pandemic has shifted more responsibility for learning to students, in line with Mahlaba's idea that we shouldn't keep students overly dependent on teachers (2020). COVID-19 has shown that teachers may not always be available as immediate sources of information, emphasizing the need to empower students to take charge of their learning. This doesn't diminish the teaching role but highlights students' involvement in constructing knowledge (Kopish & Marques, 2020). Self-directed learning, supported by various authors (Khiat, 2017; Mahlaba, 2020; Tekkol & Demirel, 2018), is a key predictor of academic success.

Additionally, the shift to technology-based teaching encourages problem-solving, critical thinking, and applied understanding (Hedding et al., 2020), as traditional assessment methods become less viable. Bester and Brand (2013) found that students possess the capability to explore new concepts, build knowledge, and link new information to real-life scenarios, facilitating effective learning in technology-rich environments. Mahaye (2020) further suggests that technology has a significant impact by enhancing access to information and elevating learning standards. This highlights that technology not only enables ongoing curriculum delivery in higher education institutions but also enhances teaching and learning effectiveness. Top of Form

The Promise of Blended and Hybrid Learning

Dziuban et al. (2004) define blended learning as a method that combines the strengths of traditional face-to-face teaching with digital enhancements from online delivery. It features student-centered learning, increased interaction opportunities, and assessment for course improvement (Watson, n.d.). Blended courses incorporate both in-person and online elements. In a blended program, e-resources are not an alternative to in-person class time; they aim to enrich and reinforce classroom concepts. Although "blended" and "hybrid" practices are often exchanged, there is a distinction: in hybrid practices, virtual content replaces in-person class time. Digital interactions in hybrid instruction can occur synchronously in real-time meetings or asynchronously, allowing students to engage at different times (Siegelman, 2019).

Hybrid and blended instruction, as noted by Singh (2017), provide students with a mix of on-site and remote learning, combining regimented and self-guided learning coursework. This flexible approach could become the new standard, particularly in disciplines where educators face challenges in delivering engaging online learning experiences, allowing instructors to reimagine and enhance content delivery, as emphasized by Rodriguez (2020). Hybrid and blended learning methods are evolving with improved technology and greater collaboration between instructors and students. They offer a safe way to learn during COVID-19 restrictions and promote digital citizenship for a flexible and secure learning environment (Akcil & Bastas, 2021).

The Flipped Classroom Model: A Student-Centered Approach:

An emerging and increasingly popular concept in education is the promotion of a child-centered strategy through the flipped classroom model. This innovative approach was pioneered by Jonathan Bergmann and Aaron Sams (2012), and initially aimed at assisting participants who missed on-site activities for various reasons. To address this challenge, they created an E-lecture series that students could access from their homes. Interestingly, these video resources began to be utilized remotely by others as well. Inspired by this process, Aaron Sams decided to incorporate this concept into their teaching and learning methodology. They began uploading videos linked to the theme before class sessions and encouraged students to view them beforehand, thereby utilizing in-class time for interactive dialogue forums and cooperative group assignments. In essence, this teaching and learning approach completely redefined the traditional classroom model (Drake, 2016).

In 2020, Lipomi endorsed video-based remote learning during the pandemic, emphasizing its benefits such as the ability for students to control video playback (Lipomi, 2020). This approach, combined with digital learning and the flipped classroom model, fosters pedagogical growth for both students and teachers. Technology integration expands learning beyond school boundaries, yielding positive educational transformations. Video, a multimedia tool merging visuals, text, and sound, proves essential and effective. Additionally, the research highlights that students prefer using critical thinking skills to engage with content through video rather than relying solely on traditional textbooks.

Alghamdi et al. (2019) identified five key factors behind the success of the flipped classroom approach: heightened student engagement, enhanced teamwork skills, increased guidance for students, facilitated classroom discussions, and expanded opportunities for teachers.

The Rise of Artificial Intelligence (AI)in Education

The success of any educational strategy or innovation hinges on its alignment with both the broader educational system and the societal context in which it is implemented. The notion that schools are not the sole source of learning and that education can occur anywhere has gained prominence, particularly during the pandemic. Within this context, the use of artificial intelligence for educational activities has been on the rise, with some studies highlighting the effectiveness of smart lesson systems in enhancing student motivation and learning (Beck et al., 1996). AI supports collaborative learning through adaptive group training, specialized assistance, virtual mediators, and virtual moderators (Luckin et al., 2016). Looking at it from the teacher's point of view, Robert Murphy highlights two crucial applications in this realm: "automated job classification systems" and "early identification of students at risk of failure or dropout" (Murphy, 2019). These applications are significant because they enable teachers to intervene promptly, offering students an efficient learning trajectory.

Augmented Reality: Reshaping the Learning Environment

The increasing effectiveness of utilizing computer applications in diverse educational contexts allows students to immerse themselves in more realistic experiences, featuring both dynamic and static images that mirror the real world. Consequently, this heightens learners' cognitive perception (Sommerauer & Müller, 2014). Additionally, AR (Augmented Reality) learning environments not only bolster learners' knowledge frameworks but also provide them with an efficient method to actively participate in these environments, ultimately achieving a greater level of proficiency (Karal & Abdusselam, 2015). Education in a simulated environment, akin to augmented reality, eliminates the need for physical relocation and has been shown to improve student learning, as demonstrated by Chen and Hsu (2016). This highlights the potential for more effective education through appropriate learning systems. AR technology, described by Cheawjindakarn et al. (2012), enhances realism by blending digital content into the real world. It's not just virtual; it provides information at the right time and place (Hughes et al., 2007). AR effectively overcomes temporal and spatial obstacles in information presentation, motivating learning and improving outcomes. This aligns with educators' dreams of ubiquitous learning, allowing continuous knowledge development by learners (Zhang et al., 2014). Billinghurst and Dunser (2012) found that incorporating Augmented Reality (AR) into educational materials, such as textbooks, enhances the learning experience. They emphasized that AR complements traditional teaching methods rather than replacing them. The study involved two groups of students with different reading abilities who studied a social studies book with AR pictures. The AR group showed more effective learning, particularly when dealing with complex concepts.

Equipping Students for a Resilient Education System

Teachers instilling resilience can reshape students' life trajectories. Although challenging, teaching resilience goes beyond traditional methods, requiring engagement and fostering self-belief. When students feel worthy and capable, they become resilient. Cahill et al. (2014) identified numerous factors contributing to resilience. These include social competence, optimism, purpose, family and school attachments, problem-solving skills, effective coping, and a positive self-image. When these factors interconnect in children's lives, resilience takes root. Furthermore, outside the classroom, participation in sports teams can impart these valuable skill sets. Resilience operates within the social fabric of everyday life, facilitating positive peer relationships, adeptly handling minor challenges, and instilling a sense of personal responsibility.

While Sarah Fuller emphasizes the changing role of teachers, it's equally vital to involve and support students in building resilient education systems. Classroom dynamics and teaching methods play a pivotal role in shaping a resilient classroom culture that can significantly impact students' lives. Traditionally, educators often relied on punitive approaches to address behavioral issues, but a new paradigm influenced by positive psychology promotes a strength-based model. This approach encourages teachers to shift their focus from students' weaknesses to their strengths while fostering overall well-being and resilience. Positive psychology aims to enable individuals to showcase their capabilities despite vulnerabilities. In the school setting, achieving this transformation involves creating a safe, psychologically secure physical environment, nurturing supportive relationships, and promoting a strong sense of belonging and positive social norms. It also entails providing opportunities for skill development and incorporating social and cultural elements from families and the community.

Conclusion

As the COVID-19 pandemic unfolded, faculty members with limited to no prior experience in online teaching were tasked with transitioning their conventional in-person classes to an online format. Instructing hybrid, blended, or online courses necessitates familiarity with best practices, effective virtual pedagogical know-how, a grasp of digital methodologies, and proficiency in web-based tools. Continuing education and training can be vital in ensuring the competence and effectiveness of faculty members new to online teaching. Thoroughly equipping instructors will facilitate the course development process and contribute to the creation of courses that ultimately enhance the learning environment for students. In the post-COVID era, as web-based and integrated learning opportunities become the norm, instructors who excel at fostering internet-based learning with a commitment to inclusiveness and building strong human relationships will significantly impact students' academic experiences. These instructors recognize that nurturing positive bonds can also motivate students to achieve their peak scholastic capacity (Pacansky-Brock et al.,

2020). Cognitive presence, defined by Garrison et al. (2000), involves constructing and confirming semantic clarity derived from a prolonged conversation in an online community of inquiry. To cultivate active cognition in remote education, faculty should provide feedback, discuss core concepts, and assess student work. Boettcher (n.d.) recommends critically evaluating students' contributions and posing challenging questions to promote thoughtful engagement in discussions. In the current and post-pandemic academic environment, technology is crucial for student learning. The digital divide, caused by limited access to technology, can hinder learning. To bridge this gap, we should increase bandwidth, create open education courses, and reconsider tuition models to provide laptops and hotspots. Additionally, implementing technology loan programs, where students can borrow devices and return them after the course, can help address this divide (Block, 2010; Brownlee, 2020).

References

Abdelhafez, A. (2021) Digitalizing teacher education and professional development during the COVID-19 pandemic, ACADEMIA, May 2021: (70). https://www.academia.edu/45095057/Digitizing_Teacher_Education_and_Professional_Development_during_the_COVID_19_Pandemic

Akcil U., Bastas M. (2021). Examination of university students' attitudes towards e-learning during the COVID-19 pandemic process and the relationship of digital citizenship. *Contemporary Educational Technology*, *13*(1), ep291. https://doi.org/10.30935/cedtech/9341

Al-Dokhny, A., & Drwish, A. (2021). Effectiveness of augmented reality in online distance learning at the time of the COVID-19 pandemic. *International Journal of Emerging Technologies in Learning* (iJET), *16*, 198–218. 10.3991/ijet.v16i09.17895

Allen, I. E., Seaman. J., & Garrett, R. (2007). *Blending. The extent and promise of blended education in the United States*, 1st edn. USA: Sloan Consortium.

Alghamdi, T., Hall, W., & Millard, D. (2019). A classification of how MOOCs are used for blended learning. In *Proceedings of the 2019 4th International Conference on Information and Education Innovations*, 1–7, New York, NY, USA.

Andrabi, T., Das, J., & Khwaja, A. I. (2008). A dime a day: The possibilities and limits of private schooling in Pakistan. *Comparative Education Review*, *52*(3), 329–355.

Arslan, G. (2021). Loneliness, college belongingness, subjective vitality, and psychological adjustment during coronavirus pandemic: Development of the College Belongingness Questionnaire. *Journal of Positive School Psychology*, *5*(1), 17–31.

Beck, J., Stern, M., & Haugsjaa, E. (1996). Applications of AI in education. *XRDS: Crossroads, The ACM Magazine for Students*, *3*(1), 11–15. https://doi.org/10.1145/332148.332153

Bergmann, J., & Sams, A. (2012). Before you flip, consider this. *Phi Delta Kappan*, *94*(2), 25.

Bester, G., & Brand, L. (2013). The effect of technology on learner attention and achievement in the classroom. *South African Journal of Education*, *33*(2), 1–15. https://doi.org/10.15700/saje.v33n2a405.

Billinghurst, M., & Dunser, A. (2012). Augmented reality in the classroom. *Computers & Education*, *45*(7), 56–63. https://doi.org/10.1109/MC.2012.111

Bliuc, A. M., Goodyear, P., & Ellis, R. A. (2007). Research focus and methodological choices in studies into students' experiences of blended learning in higher education. *The Internet and Higher Education*, *10*, 231–244.

Block, J. (2010). Distance education and the digital divide: An academic perspective. *Online Journal of Distance Learning Administration, 13*(1). https://www.westga.edu/~distance/ojdla/spring131/block131.html

Boettcher, J. V. (n.d.). *Rider university online e-coaching tips.* https://www.rider.edu/sites/default/files/files/tlc-RiderTip2ThreePresencesFS.pdf.

Briones, S. et al. (2022). Factors affecting the students' scholastic performance: A survey study. *Indonesian Journal of Educational Research and Technology, 2*(2), 97102. Available: http://ejournal.upi.edu/index.php/I JERT/

Brownlee, M. I. (2020, July 13). Here's how colleges should help close the digital divide in the COVID-Era. https://www.edsurge.com/news/2020-07-13-here-s-how-colleges-should-help-close-the-digital-divide-in-the-covid-era.

Cahill, H., Beadle, S., Farrelly, A., Forster, R., & Smith, K. (2014). Building resilience in children and young people: *A Literature Review for the Department of Education and Early Childhood Development* (DEECD). Melbourne, Victoria: Department of Education, Victoria.

Cao, W., Fang, Z., Hou, G., Han, M., Xu, X., Dong, J., & Zheng, J. (2020). The psychological impact of the COVID-19 epidemic on college students in China. *Psychiatry Research, 287*, 112934.

Castro, R. (2019). Blended learning in higher education: Trends and capabilities. *Education and Information Technologies, 24*, 2523–2546. https://doi.org/10.1007/s10639-019-09886-3

Cheawjindakarn, B., Suwannatthachote, P., & Theeraroungchaisri, A. (2012). Critical success factors for online distance learning in higher education: A review of the literature. *Creative Education, 3*, 61–66.

Chellathurai, K. G. J. (2020). Future of education post-pandemic COVID-19: Online vs classroom learning-redefining education. *The Researchers' International Research Journal, 6*(2), 23–30.

Chen, C. H., Chou, Y. Y., & Huang, C. Y. (2016). An augmented-reality-based concept map to support mobile learning for science. *The Asia-Pacific Education Researcher, 25*, 567–578.

Cherubini, L. (2020). Education in the post-pandemic era: Indigenous children and youth. *The International Indigenous Policy Journal, 11*(3), 1–11.

Davis, R., & Wong, D. (2007). Conceptualizing and measuring the optimal experience of the eLearning environment. *Decision Sciences Journal of Innovative Education, 5*(1), 97–126.

Deloitte Touche Tohmatsu India LLP. (2019). ASHE 2019 Annual Status of Higher Education of States and UTs in India. Deloitte Touche Tohmatsu India LLP, India. https://www2.deloitte.com/in/en/pages/public-sector/articles/ashe-2019.html.

DeLozier, S. J., & Rhodes, M. G. (2017). Flipped classrooms: A review of key ideas and recommendations for practice. *Educational Psychology Review, 29*(1), 141–151.

Douse, M. (2021). Let the learners lead: The worldwide transition into Post-digital age, Post-pandemic education. *Educational Planning, 28*(4), 63–75.

Drake, L., Kayser, M., & Jacobowitz, R. (2016).The flipped classroom. An approach to teaching and learning, https://www.newpaltz.edu/media/the-benjamin-center/P.Brief_2020Vision-Flipped%20classroom.pdf

Dziuban, C. D., Hartman, J. L., & Moskal, P. D. (2004). Blended learning. EDUCAUSE. *Research Bulletin, 7*, 1–12.

Eckhaus, E., & Davidovitch, N. (2019). Technology-supported teaching: technological progress or a sham? *European Journal of Educational Research, 8*(3), 697–702. https://doi.org/10.12973/eu-jer.8.3.697

Education Week, "Map: coronavirus and school closures," 2020, https://www.edweek.org/ew/section/multimedia/map-coronavirus-and-school-closures.html

Fuller, S. H. (2021) Technology as a learning space: Teachers need support for digital learning now. NORRAG Highlights. https://www.norrag.org/?blog=0&s=Fuller

Garrison, D. R., Anderson, T., & Archer, W. (2000). Critical inquiry in a text-based environment: Computer conferencing in higher education. *Internet and Higher Education, 2*(2–3), 87–105.

Goh, P. S., & Sandars, J. (2020). A vision of the use of technology in medical education after the COVID-19 pandemic. *MedEdPublish, 9*(49), 49.

Hamidi, H., & Chavoshi, A. (2018). Analysis of the essential factors for the adoption of mobile learning in higher education: A case study of students of the University of Technology. *Telematics and Informatics, 35*(4), 1053–1070.

Haningsih, S., & Rohmi, P. (2022). The pattern of hybrid learning to maintain learning effectiveness at the higher education level post-COVID-19 pandemic. *European Journal of Educational Research, 11*(1), 243–257.

Hedding, D. W., Greve, M., Breetzke, G. D., Nel, W., & Van Vuuren, B. J. (2020). COVID-19 and the academe in South Africa: Not business as usual. *South African Journal of Science, 116*(7–8), 1–3.

Hehir, E., Zeller, M., Luckhurst, J. et al. (2021). Developing student connectedness under remote learning using digital resources: A systematic review. *Education and Information Technologies, 26*, 6531–6548. https://doi.org/10.1007/s10639-021-10577-1

Hodges, C., Moore, S., Lockee, B., Trust, T., & Bond, A. (2020). The difference between emergency remote teaching and online learning. *Educause Review*, 27 March. https://er.educause.edu/articles/2020/3/the-difference-between-emergency-remote-teaching-and-online-learning. Accessed 28 May 2024.

Holmes, W., Bialik, M., & Fadel, C. (2019). Artificial intelligence in education: Promises and implications for teaching and learning, 621–653. Boston, MA: Center for Curriculum Redesign. https://discovery.ucl.ac.uk/id/eprint/10168357/

Hughes, J. E., McLeod, S., Brown, R., Maeda, Y., & Choi, J. (2007). Academic achievement and perceptions of the learning environment in virtual and traditional secondary mathematics classrooms. *American Journal of Distance Education, 21*(4), 199–214. https://doi.org/10.1080/08923640701595365

Jackaria, P. (2022). Elementary teachers' experiences and instructional challenges during the return to school after the COVID-19 closure in the Philippines. *International Research Journal of Science, Technology, Education, and Management, 2*(2), 216–225. Available: https://doi.org/10.5281/zenodo.6 955920

Jandrić, P., Knox, J., Besley, T., Ryberg, T., Suoranta, J., & Hayes, S. (2018). Postdigital science and education. *Educational Philosophy and Theory, 59*(10), 893–899. https://doi.org/10.1080/00131857.2018.1454000

Kamanetz, A. (2020, March 19). 'Panic-gogy': Teaching online classes during the coronavirus pandemic. *NPR*. Retrieved from https://www.npr.org/2020/03/19/817885991/panic-gogy-teaching-online-classesduring-the-coronavirus-pandemic

Karal, H., & Abdusselam, M. S. (2015). Increased reality. In B. Akkoyunlu, A. Isman & H. F. Odabasi (Eds.), Education technologies readings. 149–176. Ankara, Turkey: TOJET. Retrieved from https://www.rsearchgate.net/profile/Tuelin_Haslaman/publication/303880949_teknoloji_okumlari_2016/links/575a692d08aed884620d2dc2/teknolojiokumalari-2016.pdf#page=446

Kerres, M., & Buchner, J. (2022). Education after the pandemic: What we have (not) learned about learning. *Education Sciences, 12*(5), 315.

Khiat, H. (2017). Academic performance and the practice of self-directed learning: The adult student perspective. *Journal of Further and Higher Education, 41*(1), 44–59.

Kopish, M., & Marques, W. (2020). Leveraging technology to promote global citizenship in teacher education in the United States and Brazil. *Research in Social Sciences and Technology, 5*(1), 45–69.

Langthaler, M. (2021). Digitalization, Education and skills development in the Global South", NORRAG Highlights: NORRAG – Lessons from COVID 19: Digitalization calls for strong public education systems by Margarita Langthaler – NORRAG.

Laufer, M., Leiser, A., Deacon, B. et al. (2021). Digital higher education: a divider or bridge builder? Leadership perspectives on EDTECH in a COVID-19 reality. *International Journal of Educational Technology in Higher Education, 18*(51). https://doi.org/10.1186/s41239-021-00287-6

Liao, C. (2016). From interdisciplinary to transdisciplinary: An arts-integrated approach to STEAM education. Art Education, 69(6), 44–49.

Lipomi, D. J. (2020). Video for active and remote learning. *Trends in Chemistry, 2*(6), 483–485.

Liu, X., Burroughs, N. F., Tian, Q., Harvey, V. L., Housel, T. H., &; Wellingon, N. (2016). The impact of smartphone educational use on student connectedness and out-of-class involvement. *Electronic Journal of Communication, 26* (3–4).

Luckin, R., Holmes, W., Griffiths, M., & Forcier, L. B. (2016). Intelligence Unleashed: An Argument for AI in Education. Pearson. https://static.googleusercontent.com/media/edu.google.com/en//pdfs/Intelligence-Unleashed-Publication.pdf

Mahlaba, S. (2020). Reasons why self-directed learning is important in South African during the COVID-19 pandemic. *South African Journal of Higher Education, 34*(6). https://doi.org/10.20853/34-6-4192

Mahaye, N. E. (2020). The impact of COVID-19 pandemic on education: Navigating forward the pedagogy of blended learning. Available at https://www.academia.edu/42842598/The_Impact_of_COVID_19_Pandemic_on_Educ ation_Navigating_Forward_the_Pedagogy_of_Blended_Learning

Manandhar, N., & Luitel, B. (2020). Education before, during, and after pandemic: A proposal for STEAM education. Educational Souvenir: 2077, Education and Human Resource Development Center, Sanothimi, Bhaktapur, Nepal, 152–166.

Martin, F., Sun, T., & Westine, C. D. (2020). A systematic review of research on online teaching and learning from 2009 to 2018. *Computers and Education/Computers & Education, 159*, 104009. https://doi.org/10.1016/j.compedu.2020.104009

Mbhiza, H. W. (2021). Shifting paradigms: Rethinking education during and Post-COVID-19 pandemic. *Research in Social Sciences and Technology, 6*(2), 279–289. https://doi.org/10.46303/ressat.2021.31

Meirinhos, C., Fernandes, L., & Meirinhos, M. (2023). *The emergence of Artificial Intelligence in Education*. http://hdl.handle.net/10198/28816

Ministry of Human Resource Development GOI. (2019). All India Survey on Higher Education. (n.d.). Retrieved from http://www.aishe.gov.in/aishe/viewDocument.action?documentId=122

Murphy, R. F. (2019). Artificial intelligence applications to support K–12 teachers and teaching. RAND Corporation. https://doi.org/10.7249/PE315

Mushtaha, E., Dabous, S. A., Alsyouf, I., Ahmed, A., & Abdraboh, N. R. (2022). The challenges and opportunities of online learning and teaching at engineering and theoretical colleges during the pandemic. *Ain Shams Engineering Journal, 13*(6), 101770.

Muthuprasad, T., Aiswarya, S., Aditya, K. S., & Jha, G. K. (2020). Students' perception and preference for online education in India during COVID-19 Pandemic. *Social Sciences & Humanities Open, 3*(1), 100101. https://doi.org/10.1016/j.ssaho.2020.100101

Ondras, L. B., & Alvero, J. (2023). Post-pandemic challenges in addressing learning gaps: Experiences of teachers in public elementary and secondary schools. *Asian Journal of Education and Social Studies, 47*(4), 38–46.

Onyema, E. M., Ogechukwu, U., Anthonia, E. C. D., & Deborah, E. (2019). Potentials of mobile technologies in enhancing the effectiveness of inquiry-based learning approach. *International Journal of Education (IJE), 2*(01), 1–22.

Organisation for Economic Co-operation and Development [OCED]. (2020). Strengthening online learning when schools are closed: The role of families and teachers in supporting students during the COVID-19 crisis. Organisation for Economic Co-operation and Development. https://read.oecd-ilibrary.org/view/?ref=136_136615-o13x4bkowa&title=Strengtheningonline-learning-when-schools- are-closed.

Pacansky-Brock, M., Smedshammer, M., & Vincent-Layton, K. (2020). Humanizing online teaching to equitize higher education. *Current Issues in Education, 21*(2).

Palad, I. (2022). Exploring stakeholders' experiences and challenges on the printed modular distance learning modality amidst covid19 pandemic. *Jurnal Aplikasi Manajemen, 20*, 25–33. Available:https://jurnaljam.ub.ac.id/index.p hp/jam/article/view/2681

Pant, B. P., Luitel, B. C., & Shrestha, I. M. (2020, January). Incorporating STEAM pedagogy in teaching mathematics. In *Proceedings of the Eight International Conference to Review Research in Science, Technology and Mathematics Education (episteme 8), Homi Bhabha Centre for Science Education, Mumbai, India. https://episteme8. hbcse. tifr. res. in/proceedings.*

Rapanta, C. Botturi, L. Goodyear, P. Guardia, L., & Koole, M. (2020). Online Teaching University Teaching during and after COVID-19 Crisis: Refocusing Teacher Presence and Learning Activity. *Postdigital Science and Education*, *2*, 923–945. Doi: 10.1007/s42438-020-00155.

Rapanta, C., Botturi, L., Goodyear, P., Guàrdia, L., & Koole, M. (2021). Balancing technology, pedagogy and the new normal: Postpandemic challenges for higher education. *Postdigital Science and Education*, *3*, 715–742.

Reimers, F. M. (2022). Learning from a pandemic. The impact of COVID-19 on education around the world. In Reimers, F.M. (Ed.), *Primary and Secondary Education during Covid-19: Disruptions to Educational Opportunity during a Pandemic*, 1–37. Springer, Cham. https://doi.org/10.1007/978-3-030-81500-4_1

Robosa, J. et al. (2021). The experiences and challenges faced of the public school teachers amidst the COVID-19 pandemic: A phenomenological study in the Philippines. *International Journal of Advance Research and Innovative Ideas in Education*, DOI: 10.6084/m9.figshare.14028833.v1

Rodriguez A. (2020). *Do hybrid classes offer the best of both world?* University of Colorado Denver. https://news.ucdenver.edu/do-hybrid-classes-offer-the-best-of-both-worlds/.

Roy, R., & Sharma, P. (2023). Indian legal education in the post-pandemic world contextualising the impact of national education policy 2020. *Available at SSRN 4454955*.

Safonov, A. S., & Mayakovskaya, A. V. (2020). Post-digital world, pandemic and higher education. *International Journal of Higher Education, 9*(8), 90–94.

Salam, S. Jianqiu, Z. Pathan, Z. H., & Lei, W. (2017). Strategic barriers in the effective integration of ICT in the public schools of Pakistan. In *Proceedings of the 2017 International Conference on Computer Science and Artificial Intelligence*, 169–172, Jakarta, Indonesia.

Schmidt, S. M. P., & Ralph, D. L. (2016). "The flipped classroom: A twist on teaching. *Contemporary Issues in Education Research, 9*(1), 1–6.

Selvaraj, A., Radhin, V., KA, N., Benson, N., & Mathew, A. J. (2021). Effect of pandemic based online education on teaching and learning system. *International Journal of Educational Development, 85*, 102444. https://doi.org/10.1016/j.ijedudev.2021.102444

Seren, M., & Özcan, Z. E. (2021). Post pandemic education: Distance education to artificial intelligence based education: Post pandemic education. *International Journal of Curriculum and Instruction, 13*(1), 212–225.

Shulman, L. S. (1987). Knowledge and teaching: Foundations of the new reform. *Harvard Educational Review, 57*(1): 1–22. https://doi.org/10.17763/haer.57.1.j463w79r56455411

Siegelman A. (2019). *Blended, hybrid, and flipped courses: What's the difference?* https://sites.temple.edu/edvice/2019/11/05/blended-hybrid-and-flipped-courses-whats-the-difference/.

Singh J. (2017). Enhancing student success in health care programs: Active learning in a hybrid format. *Journal of Instructional Pedagogies, 18*, 1–14. https://files.eric.ed.gov/fulltext/EJ1182732.pdf.

Singh, J., Steele, K., & Singh, L. (2021). Combining the best of online and face-to-face learning: Hybrid and blended learning approach for COVID-19, post vaccine, & post-pandemic world. *Journal of Educational Technology Systems, 50*(2), 140–171.

Shadiev, R., & Sintawati, W. (2020). A review of research on intercultural learning supported by technology. *Educational Research Review, 31*, 100338.

Shafi, H. Z., Rahman, F., & Asif, M. et al. (2018). Concentrate management technologies for desalination processes leading to zero liquid discharge: Technologies, recent trends and future outlook. *Desalination and Water Treatment, 105*, 92–118.

Sharp, J., & Theiler, S. (2018). A review of psychological distress among university students: Pervasiveness, implications and potential points of intervention. *International Journal for the Advancement of Counselling, 40*(3), 193–212.

Skulmowski, A., & Rey, G. D. (2020). COVID-19 as an accelerator for digitalization at a German university: Establishing hybrid campuses in times of crisis. *Human Behavior and Emerging Technologies, 2*, 212–216.

Sommerauer, P., & Müller, O. (2014). Augmented reality in informal learning environments: A field experiment in a mathematics exhibition. *Computers & Education, 79*, 59– 68. https://doi.org/10.1016/j.compedu.2014.07.013

Sousa, D. A., & Pilecki, T. (2018). *From STEM to STEAM: Using Brain-Compatible Strategies to Integrate the Arts*, 1–264. Thousand Oaks, CA: SAGE Publications.

Sun, L., Tang, Y., & Zuo, W. (2020). Coronavirus pushes education online. *Nature Materials, 19*(6), 687–687. https://doi.org/10.1038/s41563-020-0678-8.

Tahir, I., Van Mierlo, V., Radauskas, V., Yeung, W., Tracey, A., & da Silva, R. (2022). Blended learning in a biology classroom: Pre-pandemic insights for post-pandemic instructional strategies. *FEBS Open Bio, 12*(7), 1286–1305.

Tekkol, L. A., & Demirel, M. (2018). An Investigation of Self-Directed Learning Skills of Undergraduate Students. *Frontiers in Psychology, 9*. https://doi.org/10.3389/fpsyg.2018.02324

The World Bank. (2021, March 30). Education Response and Recovery During and After COVID-19. World Bank. https://www.worldbank.org/en/topic/education/brief/mission- recovering-education-in-2021.

United Nations. (2020, August). *Policy Brief: Education during COVID-19 and beyond. United Nations.* Retrieved from https://www.un.org/development/desa/dspd/wp-content/uploads/sites/22/2020/08/sg_policy_brief_covid- 19_and_education_august_2020.pdf

Viner, R. M., Russell, S. J., Croker, H., Packer, J., Ward, J., Stansfield, C., Mytton, O., Bonell, R., & Booy, R. (2020). School closure and management practices during coronavirus outbreaks including COVID-19: A rapid systematic review. *The Lancet Child & Adolescent Health*, *4*(5), 397–404. https://doi.org/10.1016/S2352-4642(20)30095-X

Volery, T., & Lord, D. (2000). Critical success factors in online education. *International Journal of Educational Management*, *4*(5), 216–223.

Watson, J. (2008). *Blended Learning: The Convergence of Online and Face-to-Face Education. Promising Practices in Online Learning.* http://files.eric.ed.gov/fulltext/ED509636.pdf

Yang, J. (2014, May). Implementing the flipped classroom in elementary and secondary schools in China. *International Conference on Education, Language, Art and Intercultural Communication (ICELAIC-14)* (135–138). Atlantis Press.

Zhao, W., Zhang, J., Liu, X., & Jiang, Z. (2022). Application of ISO 26000 in digital education during COVID-19. *Ain Shams Engineering Journal*, *13*(3), 101630.

Zhang, J., Sung, Y. T., Hou, H. T., & Chang, K. E. (2014). The development and evaluation of an augmented reality-based armillary sphere for astronomical observation instruction. *Computers & Education*, *73*, 178–188. https://doi.org/10.1016/j.compedu.2014.01.003

Zhang, L., Basham, J. D., & Yang, S. (2020). Understanding the implementation of personalized learning: A research synthesis. *Educational Research Review*, *31*, 100339.

Chapter 17

Translation: An Efficient Tool in the Teaching of Foreign Languages

A Special Reference to the French Language

Cynthia George

Introduction

> One language sets you in a corridor for life. Two languages open every door along the way[1]

With these words of Frank Smith, we could deliberate upon

- Merits of learning a foreign language
- Drawbacks of being a monolingual

Merits of Learning a New Language

The learning of foreign languages in India has enormously increased over the past 20 years with the Third Industrial Revolution technically the Digital Revolution. The major benefits of learning a foreign language can be summed up under the following headers

DOI: 10.1201/9781003504894-17

1. Enhances communication during travel
2. Survival during stay/work/study overseas
3. Passion towards the language
4. Ensures mental stability and fitness
5. Improves their existing language skill sets by comparing one's language with that of the other
6. Excels in the job market

To Enhance Communication during Travel

> With languages, you are at home anywhere
>
> – *Edward de Waal*[2]

Let's imagine two Indian students traveling abroad to China. Let's assume that Person A knows the little basics of the Chinese language whereas Person B does not know Chinese at all. Both enter the airport, with a feel of farewell to their homeland and a series of inquisitive questions on the country of their visit lingering in their minds. Soon one is on the flight with the neighboring seat passenger being a Chinese. How would you describe the experiences of both these passengers?

Person A: (the one who knows here and there of Chinese): He tries to greet his neighbor with his broken pieces of the Mandarin dialect with of course a lot of faults. This may facilitate further exchange between the two either by way of gests; a common language may be English with a few trials on Chinese words. This may also enhance their communication further and aid this person in discovering a lot more things known only to the locals that do not appear in any books/ journals.

Person B: (the one who knows nothing of Chinese): This person too may greet his neighbor, and conversations may happen in a common language provided there is one, but still the ease of communication fails as the human brain has this tendency of moving easily from the *known to the unknown* as it generally trusts the known and doubts the unknown. Imagine a flight travel with complete silence for six hours.

Survival during Study/Stay/Work Overseas

> It is not the strongest of the species that survives, nor the most intelligent; it is the one most adaptable to change.
>
> – *Charles Darwin*[3]

It is a known fact that

- Students who move abroad for education purposes undergo a bridge course in the foreign language before commencing their actual course study. Example: Students from India who move to Russia to become doctors learn the basics of Russian.

- Apprentices/ employers are generally advised by the organizations who hire them to learn the local language quickly to facilitate communication with clients. Many organizations scrutinize resumes interviewing only candidates with the required language skill sets. The domains of hotel management, front office, marketing and logistics appreciate rather demand the knowledge of more than five to six languages from their employees
- Spending a vacation with your loved ones in the other part of the globe is quite a happy and interesting thing. Let's try to analyze the situation of a mother who has gone to assist her expecting daughter. The nuances, the small things communicated by the doctors, nurses, and medical staff have to be translated every time by the expecting daughter despite her maternity childbearing pain. On the contrary, if the mom was able to comprehend the language things would have been smooth and simple. The question arises to the point that it is the mom who takes care of the daughter or vice versa just because of this language barrier.

These are some day-to-day instances but surviving without the knowledge of the local language has/ is and will be a great handicap/ deficit.

Passion towards the Language

> A language is not just words. It's a culture, a tradition, a unification of a community, a whole history that creates what a community is. It's all embodied in a language.
>
> – *Noam Chomsky*[4]

"Hearing so much about Thiruvalluvar, I felt like learning Tamil" – a phrase from a Mauritian. The love of culture, literature, arts, etc. is one of the most important factors that contribute to the learning of a foreign language.

> Culture is one thing that automated translated technologies still can't detect. And as long as there's no way to automate culture, there will never be a way to truly automate language.

This quote helps us realize the advantages of reading/relishing the text in its original form. Furthermore, the knowledge of two or more languages aids in etymological research/understanding. It helps us to travel to the source of words/literature.

Ensures Mental Stability and Fitness

> Learning another language is not only learning different words for the same things but learning another way to think about things
>
> – *Flora Lewis*[5]

People capable of communicating in more than one language have the following cognitive benefits

a. Memory Power
b. Problem-solving/ analyzing skills
c. Critical thinking skills
d. Improved concentration
e. Multi-tasking abilities
f. Better listening skills
g. More creativity/ flexibility
h. Higher competing ability
i. Easily adaptable
j. Acceptance of one's faults/ mistakes/ failures
k. Intensifies decision-making capabilities
l. Empathy to the other

There are many other life-enhancing skills in addition to the above-mentioned cognitive benefits. These attributes also

- Reduce aging
- Improves connectivity
- Removes stereotypes/ prejudices

Improves Their Existing Language Skill Sets by Comparing One's Language with That of the Other

> He who knows no foreign languages knows nothing of his own.
>
> – *Johann Wolfgang von Goethe*[6]

An understanding of other languages helps us analyze a lot about our language. Let's consider a Tamil native learning some foreign languages and how it helps him appreciate/ understand his language

- German: The phenomenon of words compounding to form a new word is a remarkable feature of the German language. For example "Der Handschuh" means the handshoe. Similarly in Tamil வந்து + அடைந்தான் = வந்தடைந்தான். This comparison that runs in his mind helps him better learn the foreign language as it goes by the rule already quoted **known to unknown**
- French: The alphabet "J, "R" and "U" have their unique pronunciation in the French language, the Tamil student immediately compares this with his mother tongue "ழ" – the specialty of his language

Excels in the Job Market

> man who can speak two languages is worth two men.
>
> – *Napoleon Hill*[7]

Which organization would not be happy to hire one person capable of performing the task of two? As discussed in point 2, the employability factor of a person with apprehension of one or more foreign language/ regional languages is always on the higher side in comparison to a monolingual.

Advantages of a polyglot in an organization:

- The value-added profile enables smoother shifting of jobs
- Communicates with employers from different zones at ease
- Ameliorates overseas project options
- Revamps career advancements/ promotions in existing jobs

Drawbacks of Being a Monolingual

Having discussed these many gains of grasping a foreign language, let us examine the handicaps faced by a monolingual.

- In a multicultural setting, the usage of a single language leads to difficulties and resentment. The lack of understanding or minimal understanding of a language forces communicators to use simple phrases/ sentences to convey their thoughts which eventually dilute the concept that needs to be conveyed. Similarly, the hiring of a translator/ interpreter to convey the issue fails as the thought flow of the speaker/ listener is barred every time a translation/ interpretation is done. One can justify this by the example of the fate of the Latin language and its mono users in the past.
- Narrow view of the world
- Noncompatible with others/ one's faults

The list is innumerable despite some facts favoring monoglots like increased efficiency in the usage of the words of their target language. Finally one has to take a call if one on the quote

> A jack of all trades is a master of none, but oftentimes better than a master of one.

In this versatile world, it is better to be a jack of all trades and try out all new things possible, which in turn leads to self-discovery and finally to the mastery of the one that suits you.

Teaching/Learning a Foreign Language in India

With such accelerated globalization the neighboring country/ continent is available in our palms and Indians naturally polyglots due to the British colonization find it much easier to learn a foreign language than their counterparts across the planet. Thus an analysis of which method of teaching caters best to the needs of an Indian student studying a foreign language is the need of the hour.

We may apprehend this better through an algorithm on the methodology of teaching/ learning foreign languages in India over the decades namely

a. Grammar Translation Method

   Also called the traditional method, this method follows the method as to how Greek and Latin were taught in ancient times. The teaching techniques pay a great deal of attention to imparting the grammar rules of the target language. Translation of texts generally towards the target language is carried out. Teaching does not happen in the target language but in the mother tongue of the teacher/ learner.

b. Direct Method

   In complete contrast to the traditional method, the direct method emphasizes teaching the target language through the target language and completely avoids translation and grammar. This method which took form in the 1950s called itself the natural method of teaching/ learning foreign languages. Importance was given to the enrichment of the vocabulary of the target language.

c. Audio lingual Method

   This method also called the New Method or the Army Method was similar to the direct method except for the fact that the stress was on grammar and spoken aspect of the language and not on vocabulary. The teaching principle was based on repetitive oral drills and the behaviorist theory of humans

d. Structure-Global Audio-Visual (SGAV) Method

   Method developed in France in the 1970s which eliminated the teaching of linguistic rules namely grammar and syntax but promoted the holistic approach to foreign language acquisition using presenting linguistic data and by enabling physical/ effective participation of the learner.

e. The Silent Way Method

   Method developed by 20th-century Mathematician cum Educator Caleb Gategno in 1963. This method follows the submission of teaching to learning. Somewhat similar to the SGAV method, this method pays attention to the pronunciation of meaningful texts. The teacher is expected to speak the least and evaluation was done by the method of observation.

f. Suggestopedia

   A compound word formed from the words suggestion + pedagogy. This method was developed by Bulgarian psychotherapist Georgi Lozanov in 1978. This method requires a great level of skill set from the teacher as it is a combination of the Grammar Translation method of the early 20th century and the playway Way Method of the 21st century.

Teaching techniques use a pattern where

- Ample data is provided in text form with a translation
- The teacher reads the text at a comprehensible speed along with background music being played to stimulate sensory/ psychological aspects of the brain of the student
- Students repeat them at leisure and
- Finally, communicate in the target language

g. Total Physical Response Method:

One may say that this is an extension of the suggestopedia method as it stresses the kinesthetic aspect of learning. This method was framed by psychologist James Asher where teaching learning happens by physical movement of body parts. Thus the entire body is put to function in this method and all senses are kept alert all the time in this method. We may also describe this method as a comprehension approach method. One may consider Suggestopedia and TPR as the ancestors of the Play Way Method used today in our daily classrooms.

h. Immersion Method

The method primarily started in Canada with English natives learning French via French in Quebec. In this method, one teaches not only the language but also the culture in the classroom.

The immersion method can be classified into two

- Effective Immersion
- The teacher speaks the foreign language at a comprehensible speed using gestures, models, visual aids, demonstrations, etc. Thus the student can understand basic vocabulary, day-to-day conversations; cultural aspects associated with the particular text, and the nuance of the situation in the target language.
- Submersion (Ineffective Immersion)

  This method is also called the total immersion or ineffective immersion approach. The teacher speaks at a natural speed considering the student to be a native speaker. This leads to a lack of understanding or an extent only the understanding of the global concept and finally leads to distraction and loss of interest in the class among the students.

  The immersion method can be further classified based on the duration of the class

  - Total Immersion

    The language of instruction is the target language and the student spends the complete day listening, speaking, reading, and writing in the target language. Students find it difficult to understand technical, complex ideas in this method.
  - Partial Immersion

    Language of instruction is the target language but the students do not spend the entire day but only during the duration of the class during which the language is taught, the gap between classes is used by students to clarify their doubts/ discuss on the topic taught in their native language. This is a student-friendly method of teaching/learning foreign languages

- Two Way Immersion

  This is also addressed as bilingual immersion. This method is adopted in a bilingual/poly-lingual class. Instruction happens in both the target language and the language of the students. Here the student group comprises two groups; one whose native language is the target language and the other whose native language is not the target language. Thus this method increases peer group interactions and facilitates teaching/ learning within the classroom.

i. The Communicative Approach Method

This is an interactive method. The usage of authentic texts started with this method. Authentic texts are texts created for nonlanguage learning purposes but are used to teach languages.

Examples:

- Using a metro ticket to Paris, France to teach the French Language
- Using Origami Art Work of Japan to teach the Japanese Language

This method gives maximum stress and importance to the oral/ spoken aspect of the language

j. Task-Based Learning

One may call this a different version of the communicative approach, here the authentic text is a task rather than a document.

Example:

- Fixing an appointment with a doctor
- Booking a room in a hotel
- Ordering food at a restaurant

k. Computer Assisted Language Learning – CALL:

The usage of apps, programs, and MOOCs all fall under this category of learning. When face-to-face learning combines with CALL we would end up with blended learning – the strategy widely accepted and used during this pandemic across the globe.

Having discussed various teaching-learning methodologies and their mutations over the centuries one has to accept that there is no hard and fast rule to the teaching/ learning of foreign languages but it is for the teacher to decide which method works out best for his/ her student.

Teaching/Learning of French in India

Language is a theater in which words are the actors

– *Ferdinand Brunetière*[8]

If so, India will be dealing with thousands of theaters and millions of actors. According to the most recent survey, there are 1652 languages in India, 22 of which have been made official by the Constitution, with English providing communication between the various language groups in the country. French is nowadays slowly establishing itself as the most important foreign language in educational circles with around 500,000 learners and 6,000 teachers. Teaching/ Learning of French like any other foreign language has undergone various transfigurations over the years. Can any teacher sum up that one particular method of teaching is sufficient in a poly-lingual, heterogenic country like India? The answer is NO.

French is taught in India at

- Alliance Française (AF)
- Private Institutes
- Universities
- Colleges
- Schools

An analysis of the

- Syllabus
- Evaluative System
- Recruitment of teachers
- Motivation of students
- Location of the institution
- Currently used teaching methodology

would help us find the most efficient teaching methodology for the blooming foreign language French

Teaching/Learning at the Alliance Française

I would like to separate teaching/ learning at the Alliance Française from that of the teaching/ learning happening at universities/ schools and colleges. The parameters of analysis mentioned above completely vary between AF and the other Indian Institutions.

Syllabus

AF follows the CEFR - The Common European Framework of Reference for Languages based on European research and stimulates the communicative approach method combined with the immersion method of teaching/ learning. The students are classified as 6-level users based on his/her oral and written comprehension and production respectively.

Evaluative System

DELF – (Diplôme d'Etudes en Langue Française) and DALF (Diplôme Approfondi de Langue Française) are the French diplomas offered by the Ministry of Education, France. The evaluative system evaluates the following components namely

- Oral comprehension
- Oral Production
- Written comprehension
- Written Production

Recruitment of Teachers

Teachers are recruited periodically to ensure that the student-staff ratio never exceeds 1:20. Teachers also undergo formation courses from France.

Motivation of the Students

Students joining in a French course at the alliance have a considerable desire to learn the language. Paying a fee of around Rs. 10,000 ensures the participation of the student in the interactive class

Location of the Institution

With world-class infrastructure AF enables blended learning with maximum usage of CALL, Air conditioned classrooms with smart classes are the norms of the day at the AF. Another remarkable fact is that AF is located in urbanized locations and not remote interior villages of India

Currently Used Teaching Methodology

The submersion method accompanied by CALL ensures communicative/ interactive teaching-learning.

Teaching/Learning at Indian Institutions

Now let us analyze the same parameters in an Indian context with teaching/learning French in Indian Institutions

Syllabus

The syllabus taught in schools and colleges ensures that French is taught from the basic (alphabet) and is considered light and easy in comparison to the heavy burdensome syllabus of the counter native languages which demand learning of literary texts. Only at the master level, one start learning hard-core French literature/ civilization

Evaluative System

All languages are to be tested for the four competencies - listening, reading, speaking, and writing. The evaluation system here in India gives more value or almost all the importance to the written aspect of the language.

Example: The end semester examinations bear 75% of the total marks and this is entirely evaluated based on the written component of the language.

Recruitment of Teachers

The difference between the number of qualified French teachers and the number of institutions offering French is vast. This lack in the number of qualified French teachers

- touches on the recruitment of part-time teachers
- pushes teachers to work in various institutes
- causes the recruitment of underqualified teachers

Since the number of part-time teachers and those employed in several institutions is significantly high, there is a time constraint for the teacher to complete his program. This time constraint forces the teacher to focus only on the part that is being evaluated - the written aspect. Thus the unevaluated competencies are thrown for a toss.

As there is a demand for French teachers, teachers' salaries are high and some institutions do not want to pay them; which leads to the other problem of recruiting underqualified teachers (this is not a generalization, but a notable fact in some colleges).

Motivation of the Students

Only a handful of students have a real passion for learning a foreign language because, as we saw in the beginning, most of them choose French to avoid the great syllabus of other languages offered in their colleges.

Location of the Institution

Not all Universities/ colleges/ schools in India are located in urban sophisticated cities; we have schools/ colleges even in tribal hilly villages which offer French as a subject to their students.

Currently Used Teaching Methodology

Despite the above-mentioned challenges French teachers across the country facilitate learning through innovative strategies with available resources namely

Teaching of French through

- Chalk and Talk Method
- Demonstrations
- Play Way Method
- Music
- Theatre
- Poetry
- Cinema
- Guest Lectures
- Field Visits

In all these methods in the Indian context with the above-cited restrictions, the complete immersion method cannot be applied in day-to-day teaching/ learning of French.

Role of Translation in the Teaching of French in India

Traditional methods used translation as an important tool in teaching French as a foreign language. Over time, this use of translation in language teaching has been rejected as modern and recent methods emphasize only the oral competence of communication. But in an Indian context, one cannot deny two important facts

- Usage of bilingual immersion – the student-friendly method as mentioned before to facilitate teaching learning in a heterogeneous class with natural poly-linguals, the Indians
- The evaluative system which compromises 30% exercises translation.

We do not advocate that every letter, every word, and every sentence needs to be translated into French, neither does one argue that grammar rules have to be taught explicitly but French can be taught via English or the mother tongue of the student.

Advantages of using translation in the teaching of French as a foreign language

- Facilitates comprehension
- Learning happens from the known to the unknown
- Inductive method of teaching
- Best suited for a heterogeneous group of students
- Favors polylingual
- Natural method of learning

Facilitates Comprehension

Translation helps the student to understand the target language vocabulary with more confidence, it also aids in clarification of doubts/ queries that appear in the target language.

Learning happens from the known to the unknown:

As mentioned earlier known is a comfort zone as learning in the comfort zone puts the brain at ease; any learning is made easy when the student enjoys it. The immersion method creates a feeling of insecurity as it is completely foreign and unknown but this bilingual immersion or usage of translation in class creates trust among students and accelerates teaching/ learning.

Inductive Method of Teaching

With the known language being the medium of instruction teaching and learning happen from

- Simple to complex
- Concrete to abstract
- Analysis to Synthesis
- Particular to general
- Empirical to rational
- Psychological to logical Actual to Representative
- Whole to Parts
- Definite to indefinite

Best Suited for a Heterogeneous Group of Students

Let's take the example of a Class VI French class with students whose mother tongues are Hindi, Tamil, Telugu, Kannada, Malayalam,.....The moment the teacher uses the word "Je" the students translate it into their respective mother tongues and comprehend the word better

Hindi – मैं
Tamil – நான்

Telugu – నేను
Kannada – ನಾನು
Malayalam – ഞാൻ

Interested students may also call out the words in their languages enabling peer group discussions within a heterogeneous class.

Favors Polylingual

Indians are naturally capable of speaking more than one language; the British rule has enabled comprehension of English along with their mother tongue. It is a scientifically proven fact that learning a third language is easier than a second language; also the brain uses the vocabulary of these two known languages to learn the third foreign language. Then imagine Indians who are already capable of speaking multiple regional languages, learning a foreign language would just be a cakewalk.

Natural Method of Learning

Translation is the natural way of comprehending a new language. Imagine you are on a telephone call and the receiver does not understand the word, the first natural thing that one does is better articulation and increase the sound of the voice, still if comprehension is not experienced, one instantaneously uses a different word in a different language to bring sense to the conversation. So why avoid natural oxygen and give importance to inhalers?

Conclusion

I conclude with the words of Nelson Mandela

> If you talk to a man in a language he understands, that goes to his head.
> If you talk to him in his language that goes to his heart[9]

And if any teaching reaches the hearts of the students that produces better learning than the ones reaching the brain.

By the words of Burns R

> There are no two learners who progress at the same speed. No two learners are ready to learn at the same time. There are no two learners who use the same study techniques. There are no two learners who solve problems in the same way. There are no two learners who are motivated to achieve the same goals. The teacher is thus responsible for finding the needs of his student and responding to him[10]

Thus there is no single strategy in the teaching of a language but based on the needs of the student the teacher has to find multiple strategies to aid his/ her students

References

1. https://www.goodreads.com/quotes/1216027-one-language-sets-you-in-a-corridor-for-life-two
2. https://www.goodreads.com/quotes/768847-with-languages-you-can-move-from-on e-social-situation-to
3. https://quoteinvestigator.com/2014/05/04/adapt/
4. https://multilingual.com/articles/community-lives-creating-a-universal-language/#:~:text=According%20to%20linguist%20and%20political,all%20embodied%20in%20a%20language.%E2%80%9D
5. https://yolainebodin.com/the-language-nook/quotes/quote-about-learning-another-language-by-flora-lewis#:~:text=%E2%80%9CLearning%20another%20language%20is%20not,who%20specialised%20in%20international%20affairs
6. https://laacademia.co.uk/he-who-knows-no-foreign-language-knows-nothing-of-his-mother-tongue/
7. https://www.englishclub.com/ref/esl/Quotes/Language/A_man_who_can_speak_two_languages_is_worth_two_men._2919.php
8. https://bonjourfrenchwords.tumblr.com/post/141870245383/language-is-a-theater-of-which-words-are-the/amp
9. https://www.brainyquote.com/quotes/nelson_mandela_121685
10. http://www.ac-grenoble.fr/ien.aix-les-bains/wp-content/uploads/2015/04/7__postulats_de_Burns-Marseille.pdf

Chapter 18

Immediate Effect and Future Perspective of COVID-19 Pandemic on School Education

Tayum Saroh

Introduction

The COVID-19 pandemic has brought the entire world to a standstill due to the lockdown. COVID-19 virus has spread rapidly into every corner of the world and brought untold misery to the entire humanity (Wikipedia, n.d.). COVID-19 pandemic has badly affected the education sector of most of the nation, which is a critical determinant of a country's economic future. The COVID-19 virus was first identified the Wuhan, China on December 31, 2019 and it was declared a pandemic by WHO in 2020. In India, the first case of the COVID-19 pandemic was reported in the state of Kerala on 30 January 2020 (Wikipedia). Since then the COVID-19 virus has infected more than 3 crore people in India during the first and second waves of the pandemic (Wikipedia, n.d.). The number of infected people is increasing daily and many have lost their lives. The second wave of COVID-19 was witnessed from April 2021 with the infection surging over 3 lakh fresh cases per day. The increase rate of infection continued to increase in May in which more than 4 lakh fresh cases daily were also reported. The healthcare workers work day and night to take care of the infected serious patient and many of them die in the line of duty. The government also tried its level best to control the people from getting

 DOI: 10.1201/9781003504894-18

infected with the virus by imposing lockdown and creating awareness through Radio, TV, Newspapers, posters, banners, and various social network platforms (Burgess & Sievertsen, 2020).

The Government of India has initiated several preventive measures to control the spread of COVID-19. The central government has closed down all the educational institutions of the nation from 16th March 2020. The CBSE also postponed class X and XII examinations all over the nation and issued guidelines to all the examination centers to maintain COVID-19 SOP. Similarly, most of the educational boards of the country and the state governments also postponed the examinations due to the spread of COVID-19 virus. On March 22 the country observed a nationwide 'Janta Curfew' and the government imposed Lockdown from March 25, 2020, onwards. The government has wisely extended the lockdown, due to the increase in COVID-19 cases in the country. The educational institutions remain closed during the entire lockdown period. The educational institutions were partially opened by maintaining strict SOP from December 2020 to mid-March 2021 just before the arrival of the second wave of April 2021. The CBSE has canceled the Class X and XII examinations for the academic session 2020–2021 and all the educational institutions were closed down once again due to the arrival of the second wave of the pandemic. The government has imposed a lockdown and started a vaccination drive to control the further spread of COVID-19 virus. It is still unclear for how long COVID-19 pandemic will continue to haunt the entire country of the world as it always comes up with new variants before we understand the behavior of the existing variants of the virus.

Government Initiatives to Provide Education during Pandemic

The pandemic has taught teachers, students, and parents to look for alternative methods of traditional schooling since the complete shutdown of schools starting from the first phase of nationwide lockdown in March 2020. The second wave of the pandemic brought deadlier variants of the virus and the record numbers of virus infection was recorded in April and May 2021. The educational institutions are yet to open for offline classes. It is worth mentioning here that the government has instructed online classes to all educational institution to ensure academic activities going on during the lockdown period during the first and second waves of the pandemic. It was due to the lockdown that digital technology was well explored by the students and teachers alike. The Ministry of Human Resource Development (MHRD) has made several arrangements, including online portals and educational channels through direct-to-home TV and radios for students to continue learning (MHRD notice, 2020). The digital initiatives of MHRD for school education during COVID-19 are listed below

Diksha Portal

The Diksha portal is a program initiated by the government, which was prepared under the supervision of CBSE and NCERT by more than 250 teachers capable of teaching in multiple languages to provide e-learning materials for the students, teachers, and parents in parallel to the curriculum. In the portal, many learning materials are available in the form of worksheets, video lessons, textbooks, and assessments to check students' progress. The learning materials are available for Classes -I–XII in different languages. The Diksha app could be downloaded online from the Google Play Store and IOS.

e-Pathshala

It is an e-learning app created by NCERT for classes 1 to 12 in multiple languages. The e-Pathshala provides stocks of videos, books, audio, etc., materials for the teachers, students, and parents to help them learn in different languages. It contains almost 2,000 videos, 1,886 audios, 696 e-books, and 504 Flip Books for classes 1–12 in multiple languages. The stakeholders can download the Mobile Apps of e-Pathshala online and take advantage of the materials available in the portal (Jena, 2020).

National Repository of Open Educational Resources (NROER)

The National Repository of Open Educational Resources portal provides facilitates teachers and students to learn online in multiple languages by providing various resources in the form of interactive modules, books, educational videos, and a host of STEM-based games. The content of NROER is plotted with the existing curriculum of classes I–XII which is added with study materials for the teachers. The portal has It has a total of 14,527 files including 401 collections, 2,779 documents, 1,345 interactive, 1,664 audios, 2,586 images, and 6,153 videos in multiple languages. Interested learners can visit the portal to avail the facilities provided in the portal.

Immediate Effect of COVID-19 Pandemic

The Schools Are Shutting Down

The COVID-19 pandemic has resulted in the shutting down of all the educational institutions in the nation (Emgzell et al., 2020). The physical classes are highly affected by the shutting down of institutions and the second wave that hit the nation during May and June 2021 have worsened the hope of opening school sooner or later. Online classes for higher and school education started gradually after the lockdown was imposed in the entire nation in May 2020. It was assumed that the classes would be started offline from the academic session 2021–2022 and many parents did not enroll their children in the school in the hope of opening the school. Both the central and state governments do not want to take the risk of opening schools after assessing the impact of the second wave of a pandemic on the health of the citizens.

Online Education

The schools are directed by the government to continue to educate the children through online mode. The online class is the best option for children to continue their education without taking the risk of exposing themselves to viruses. However, it is also equally true that many parents are not satisfied with the online classes of their children due to their less exposure to online classes. Some also believe that the sudden shift from physical classes to online learning will not bring any fruitful results for the children. It is only through the online classes the teaching-learning process is continuing among the children during this pandemic era. The parents and the children who have already learned about the use of the right technology are getting lots of advantages from online classes. Some research shows that on average, students *retain 25–60%* more material when learning online compared to only 8–10% in a classroom. This is mostly due to the students being able to learn faster online; e-learning requires *40–60% less time to learn* than in a traditional classroom setting because students can learn at their own pace, going back and re-reading, skipping, or accelerating through concepts as they choose (Li & Lalani, 2020).

Emerging Online Education Platform

The pandemic also allowed various online platform and social network to hold their foot in the market of online education. Various online platforms such as WhatsApp, YouTube, Google Classroom, Cisco Webex, Google Meet, and many more are used by various educational institutions and teachers to get connected with their students, parents, friends, and colleagues. These online platforms were emerging as the alternative to online education and physical gatherings for people during this pandemic period. It is also attracting people who are not familiar with smartphones and social networks as an alternate way of keeping in touch with friends.

Depriving of Education to Rural Masses

The rural masses of the nation are frequently facing network and communication problems, especially during monsoon due to rain which it very difficult for them to attend online classes. The government has established many schools in the interior parts of the country after the implementation of Sarva Shiksha Abhiyan (SSA) to provide education to the walkable distance of the school-going children. The shutdown of schools due to COVID-19 pandemic has once again made them suffer to continue their learning during the pandemic time. They are facing multiple problems in attending online classes. On one hand, it is very difficult for them to attend online classes due to network issues, and on the other hand purchasing and handling Smartphones is very difficult for them.

Changing Role of Teacher

The role of the teacher also gets changed due to the online classes. The teachers normally look after the students and physically share emotional bonds with the students along with providing education to the children in the school. The online classes make the teaching-learning process more fulfilling the formalities for many teachers and students. One of the main behind this happening is the sudden shift from physical classes to online classes. The students and the teachers both take time to adjust mentally and physically to face this paradigm shift in the education process (Mahanand, 2020). The teacher can handle the students to learn in front of them during the physical class, which is not happening in online classes due to physical distance.

The burden of Child Education Shifted to the Parents

The children are now staying in their homes to pursue education due to online classes. The teachers were mostly responsible for educating the children at school during physical classes and sharing the burden with the parents to look after them during school time. However, due to shut down of schools parents are bound to purchase either Smartphones or tablets to continue the education of their children through online classes. The parents are also bound to present with them during online classes to clear their doubts. So, most of the burden to educate the child is now shifted to the parents.

Future Perspective of School Education after COVID-19 Pandemic

Concept of Online Educational Institution

The COVID-19 pandemic lockdown has given us a chance to explore the concept of online educational institutions to a higher extent. The ODL system has already existed and functioned very successfully in the educational system since long ago. The emerging demand for online classes during the nationwide lockdown has brought more curiosity to explore the various areas and programs in which the education system could be made accessible to all during any kind of crisis by maintaining quality and quantity. The government through different interventions already initiates the concept of an online education system (Study Abroad Life, 2020). The day is not far when all the physical modes of education will be stopped to give way to online education.

Preparedness to Face Challenges

The present situation of lockdown and suffering due to the pandemic has brought a lot of misery to people of the entire nations of the world. At the same time, it gives all of us a lesson to be more prepared to face the upcoming challenges of the future. It was due to the pandemic many scientists, academicians, sociologists, psychologists, etc. are doing lots of experiments and exploring to cope with the sudden unique situation. The outcome of the research and exploration will be very helpful for the entire upcoming generations to face new challenges in the future.

Challenging Role of Teacher

The role of the teacher is becoming more challenging due to the change in the entire system of education. The sudden paradigm shift in curriculum has made the teacher adjust the changing curriculum to fit into the new system of education. The present situation may become normal after some time but the impact of the current pandemic on the system of education will last for a long period. The teacher has to prepare himself or herself to face changes that are going to occur in the field of education due to social change that will happen after a pandemic.

Changing in Teaching Methods

The present methods of teaching applied by a teacher may be badly affected by the demand for education after COVID-19 pandemic lockdown. The children and the parents are now aware of the problems they have faced during this pandemic to continue their education. The younger generation may demand a more suitable system of instruction from the teacher to suit them to adjust when this kind of crisis occurs.

Change of Curriculum

The term 'social distancing' is becoming very popular nowadays and many people are bound to follow this term in their daily lives. The new generation is already stricken by the concept of the nuclear family and the system of joint family is becoming very weak. There is a possibility that the suggestions to maintain appropriate behavior to avoid the spread of the pandemic might have a high impact on the mentality of the upcoming generation, which may trigger a change in the social system. The social change demands a change in the entire curriculum of education to fulfill its needs.

Flexible Curriculum

The present system of education suggests a flexible curriculum for better learning and enjoyment of education by the children especially at the lower level. The shutting down of educational institutions has opened for the blended mode of learning which invites more flexibility in the entire education system (UNESCO, n.d.). If the curriculum becomes more flexible the students will learn at self-pacing, choosing the stream of their choice, maintaining flexibility in learning time, and many more. The flexible curriculum will also be child-centered and change the mode of assessment to make it less rigid and give more opportunity to the students to enjoy the learning.

Conclusion

The COVID-19 pandemic has brought a drastic change in the education system of India from traditional schooling to a digital-based education. It has affected the education of all sections of society and brought untold miseries. Many students were deprived of school education due to the sudden change in the system of education to a digital based. The government has also taken various initiatives to provide digital education to all the school-going children of both urban and rural areas. The sudden impact of the pandemic hurt the lives of both students and parents both externally and internally. Many rural students are still deprived of school education due to poor mobile networks and communication problems. Although the pandemic has brought a drastic change in the education system some positive perspectives in the field of education have also been witnessed in this pandemic era. Our young generations are getting ready for the future digital-oriented environment. The concept of blended learning is also emerging as a good option to face new challenges in the future. This is the right time for an academician, scientist, philosopher, and engineer to ponder and explore new things and ideas to face the same type of problems in the future.

References

Burgess, S. and Sievertsen, H. (2020). Schools, skills, and learning: The impact of COVID-19 on education, BOXEU. Retrieved from https://voxeu.org/article/impact-covid-19-education

Emgzell, P., Frey, A. and Verhagen (2020). Learning loss due to school closures during the COVID-19 pandemic, Proceeding of National Academy of Sciences for the United States of America. Retrieved from https://www.pnas.org/content/118/17/e2022376118#ref-5

Jena, P.K. (2020). Impact of pandemic COVID-19 on education in India. *International Journal of Current Research* 12(7). Retrieved from https://www.researchgate.net/publication/342123463_Impact_of_Pandemic_COVID-19_on_Education_in_India

Li, C. and lalani, F. (2020).The COVID-19 pandemic has changed education forever. This is how the World Economic, Forum. Retrieved from https://www.weforum.org/agenda/2020/04/coronavirus-education-global-covid19-online-digital-learning/

Mahanand, S. (2020). Impact of covid-19 on education and education sectors, *Education Asia.in*. Retrieved from: https://educationasia.in/article/the-impact-of-covid-19-on-education-and-education-sectors-know-here

MHRD notice (20 March, 2020). COVID-19 Stay Safe: Digital initiatives. Retrieved from https://www.mohfw.gov.in/pdf/Covid19.pdf Retrieved from https://en.wikipedia.org/wiki/COVID19_pandemic_in_India

Study Abroad Life (2020). How Covid-19 will affect the Indian education system. Retrieved from https://www.studyabroadlife.org/how-covid-19- will-affect-the-Indian-education-system/

UNESCO (n.d.). COVID-19 Educational disruption and response. Retrieved on WHO. WHO Coronavirus Disease (COVID-19) Dashboard. Retrieved from https://covid19.who.int/ Wikipedia. Covid-19 Pandemic in India. Retrieved from https://en.wikipedia.org/wiki/Education_in_India

Wikipedia (n.d.). Education in India. Retrieved from https://en.unesco.org/covid19/ education response

Chapter 19

Multilingualism in Global Education and Language Learning

A. Sivakumar and V. Vanitha

Objectives

1. To illuminate the multifaceted cognitive advantages that multilingualism confers upon individuals.
2. To explore the dynamic and ever-evolving landscape within the realm of global education.
3. To examine the intricate role that multilingualism plays in nurturing and enhancing the rich tapestry of cultural diversity.
4. Delving deep into the nuanced concepts that underpin both multilingualism and plurilingualism.
5. To undertake a thorough and intricate analysis of language policy, meticulously scrutinizing its manifold impacts on the complex phenomenon of multilingualism.
6. To conduct a profound assessment of multilingualism as a quintessential global competency, positioned within the intricate dynamics of our increasingly interconnected and globalized world.

Chapter Organization

The upcoming chapter comprises the subsequent divisions:

1. Introduction

DOI: 10.1201/9781003504894-19

2. Multilingualism: A Cognitive Advantage
3. The Evolving Landscape of Global Education
4. Multilingualism and Cultural Diversity
5. Multilingualism and Plurilingualism: Understanding Linguistic Diversity
6. Balancing Global Competence and Native Language
7. Language Learning and Identity
8. Language Policy and Multilingualism
9. Language and Globalization
10. Language in the Globalized Economy: Shaping Communication and Competency
11. Multilingualism and Knowledge Acquisition
12. Promoting Multilingualism: A Societal Imperative
13. Preparing for a Globalized World
14. Conclusions
15. References

Introduction

Language serves as a means of expressing one's thoughts and feelings as an individual, reflecting the values and continuity of culture within society, and functioning as a communication tool. It plays a pivotal role in various cognitive and social aspects, including acquiring and generating knowledge. Research in child development and language acquisition indicates that young children acquire literacy skills most effectively when they engage with their "local language," spoken in their homes, both as a language of expression and as a medium of instruction. Furthermore, children aged 2– 8 exhibit remarkable flexibility in learning multiple languages, a crucial social skill that should be nurtured, in addition to the well-documented cognitive advantages of being multilingual.

Understanding the intricate functions of linguistics and knowledge transactions in the educational province is of paramount importance. This issue has long captivated the attention of the realm of English speakers. Bullock Report (1975), aptly titled "A Language for Life," emphasized and declared that language is the linchpin for expanding knowledge throughout the entire curriculum. Language, as the report asserted, played an indispensable role in all subject areas, serving both as the medium to impart knowledge and the tool to acquire understanding. Consequently, the Bullock Report boldly stated that "every teacher is a teacher of language." This report had a pervasive outcome on educators and policymakers in countries where English is the primary language.

Academic facilitators like Douglas Barnes (England), Jerome Bruner (United States), and Jim Cummins (Canada), among others, elaborated on the influence of the Bullock Report. They collectively argued that learning and communication were inextricably linked, positing that knowledge and language were indivisible entities, a concept succinctly captured by the notion that "there can be no learning without communication, and our knowledge is inseparable from the language we use to express, store, and access it" (Little, in King & Carson 2010).

In today's diverse classrooms, a crucial distinction lies in the differentiation between "action knowledge," representing the information that students possess before entering school, and "school knowledge," which is characterized by the precise and theoretical language employed within the curriculum. Douglas Barnes, in addressing the issue of school learning, aptly noted that students might grasp this knowledge to some extent (Barnes, 1976), enough to respond to a teacher's questions, yet it remains somewhat distant, akin to someone else's knowledge rather than their own. Barnes astutely pointed out that without the incorporation of "school knowledge" into their worldview, students would only have a partial understanding, hindering their progress. This presents a substantial challenge for educators.

Jim Cummins (2013), in Canada, further developed this distinction by contrasting "BICS" (Basic Interpersonal Communication Skills) and "CALP" (Cognitive Academic Language Proficiency), subsequently pared down to "conversational" and "academic" language proficiency. "Conversational" language represents the unofficial dialogue applied beyond the confines of the classroom, is hugely context-dependent, and serves as the initial way students perceive and interact with the world, instinctively assimilating various languages.

The conventional and commonsensical perspective on language policy, often termed "language planning," typically involves a clear demarcation of different facets of language. Various characterizations of language policies worldwide, as exemplified by LeClerc in 1999, encompass a range of approaches such as "assimilation," "non-intervention," and "bilingualism." Interestingly, it's rare to find genuinely all-encompassing language policies at the state level. In its place, there are typically separate policies about the official national language, the teaching of foreign languages, and the encouragement or oversight of minority languages.

In the contemporary political landscape of the 21st century, language seems to be conceptualized as a collection of distinct yet not always interconnected phenomena. This categorization includes the national language, second languages, foreign languages, and minority languages, each governed by a separate policy framework.

Every individual possesses the fundamental right to express themselves in a language of their choice on the Internet and in public spaces. This right is firmly established in International Human Rights agreements, such as the Universal Declaration of Human Rights, and is underscored by the UNESCO Recommendation on the Promotion and Utilization of Multilingualism and Cyberspace Access from 2003. This unique normative framework encourages collaboration among international bodies, governments, civil society, academia, and private sector entities, including the IT industry. Its objectives encompass the development of content and systems that support multiple languages, making it easier to connect with networks and access systems, producing content for the realm accessible to the general public without legal restrictions., and aiming to strike an equilibrium between the concerns of copyright holders and the wider public interest.

Multilingualism: A Cognitive Advantage

Multilingualism is imperative in India, as it is in many developed nations, and should be viewed as a blessing and a chance for personal growth rather than a burden. When children are exposed to multiple languages from an early age, they can acquire them rapidly. Research worldwide consistently demonstrates that multilingual children tend to excel in their learning and prospects compared to their monolingual counterparts. This enriches them both intellectually and culturally, allowing them to think in multiple ways throughout their lives. They possess the linguistic frameworks, lexicon, expressions, and subtle literary elements of multiple languages, constituting a valuable resource. Introducing the three-language formula in the curriculum showcases the importance of languages to survive in the technical world.

The National Education Policy of India in its 2020 framework has given special importance to language education and has created a section for multilingualism where it emphasizes nurturing and embracing diverse linguistics within the educational framework. The policy also states that language is the most competent tool for developing the cognitive and affective domains of the students. By incorporating multiple languages, we can create a more inclusive environment and also provide vast educational and employment opportunities to the students.

India boasts a linguistic diversity that is among the richest, most scientific, most aesthetically pleasing, and most eloquent globally. It possesses an extensive repository of both ancient and contemporary literature, encompassing prose and poetry, as well as a rich heritage of film and music composed in these languages. These linguistic assets not only contribute to India's national identity but also constitute a substantial cultural and intellectual wealth.

The Evolving Landscape of Global Education

Globalization has significantly altered the landscape of education today. The rapid advancement of technology and the exploration of changes have reshaped societies worldwide, leading them to become more multicultural and multilingual. Consequently, there is a growing demand for holistic education that can cater to the diverse needs of these evolving societies.

Language functions as a cohesive medium for communication, and possessing expertise in numerous languages greatly enhances the acquisition of diverse subjects. Contemporary pedagogy in the 21st century places a strong emphasis on classrooms that are immersed in language and the cultivation of adept communication skills in students. To enrich foundational knowledge, encourage innovation, and foster cooperation among various domains, innovative educators have spearheaded an effort to integrate disciplines such as the humanities, language arts, performing arts (dance and drama), music, visual arts, design, cultural studies, and new media with the STEM subjects (Science, Technology, Engineering, and Mathematics).

The earliest known advocate for this integrated educational approach was Friedrich Froebel, a renowned German educator credited with inventing Kindergarten, who paved the groundwork for what we now recognize as STEAM long before the acronym existed, around 1830. Later, in 2006, Georgette Yakman formulated the STEAM educational framework, which emphasizes the interconnectedness of various subjects and their relevance to a rapidly changing world, thereby fostering the development of skills applicable throughout one's life (Yakman, 2006).

As Leeuwen, Vovers, and Silk (2018) highlight, the significance of STEAM education extends beyond just the STEM subjects, as it represents a global movement with the capacity to potentially instigate a significant transformation in the methodologies of educational methods and acquire knowledge in the contemporary era.

Multilingualism and Cultural Diversity

In today's rapidly changing multicultural and multilingual society, there is a pressing need for a holistic approach to education. Holistic education aims to nurture and encourage the growth or advancement of every individual's innovative, communal, cognitive, physical, emotional, artistic, and spiritual capacities. It goes beyond traditional teaching methods and seeks to engage students actively in acquiring knowledge, promoting a sense of personal and collective responsibility among them.

The National Center for Biotechnology Information (NCBI), a division of the National Institutes of Health (NIH) in the United States, has provided evidence supporting the idea that polyglot kids enjoy several intellectual, affective, and cultural benefits compared to their monoglot counterparts. In a country like India, where enhancements in educational accessibility frequently entail a potential loss of one's native language due to a disconnect between the language of instruction in schools and the household language often differ, multilingual education often assumes a pivotal role.

This strategy helps strike a balance between the necessity of a global corporate language, such as English, and the advantages of learning in one's mother tongue. It is widely recognized that the employees of the future must possess a range of competencies and, at the very least, an appreciation for how various skills and cultures can complement one another. This recognition is essential for preparing individuals to thrive in an increasingly interconnected and diverse world.

Within the field of education, the primary objectives often revolve around enhancing one's ability to communicate in multiple languages and nurturing a deeper understanding of different cultures as the primary benefits of language acquisition.

Multilingualism and Plurilingualism: Understanding Linguistic Diversity

Multilingualism pertains to the coexistence of multiple language varieties within a specific geographic region, regardless of whether these varieties are officially recognized as distinct languages. In such areas, individuals may be monolingual, meaning they speak only one language.

On the other hand, plurilingualism refers to the linguistic repertoire of individuals, encompassing multiple language varieties. This concept stands in contrast to monolingualism and includes a person's "mother tongue" or "first language" in addition to any extra languages or different language forms they are proficient in. Therefore, within multilingual areas, you may find a mix of monolingual individuals, who speak only one variety, and plurilingual individuals, who have command over multiple language varieties.

In certain regions, the primary linguistic focus may revolve around preparing for globalization by enhancing the plurilingual competence of the local population and creating a more welcoming environment for visitors, whether they are involved in trade or tourism. In such cases, policy initiatives tend to prioritize language education in schools and universities, particularly concerning the major languages used in international communication. Additionally, efforts are directed toward providing a variety of languages and resources to assist tourists.

Notably, many of these places often opt to adopt English as the acknowledged 'preeminent' language. The linguistic mode spoken by people who have come to live in a new country, often from another nation may often go unnoticed, despite their presence. In several places, particularly those that were historically cities where people were leaving rather than arriving until recently, there might be no organized approach to addressing the linguistic difficulties presented by immigration or catering to the requirements of new arrivals. This is because some people view immigration as a transient occurrence at times that may eventually subside. Numerous cities have transitioned from having growing immigrant communities to seeing residents leaving for other places during economic downturns and then back to net immigration during periods of economic prosperity (e.g., Buenos Aires, São Paulo).

Examining Globalization from a Linguistic Standpoint

According to Johnson (2001), in an increasingly globalized world, languages are no longer confined to specific geographic regions or individual nation-states. They, like "goods, capital, people, knowledge, images, communications, crime, culture, pollutants, drugs, fashions, and beliefs," travel across global boundaries (McGrew, 1992, p. 65). Globalization is exerting profound effects on languages, and to comprehend globalization from a linguistic viewpoint, it becomes crucial to examine language deployment and significance within a universal backdrop. This perspective, as Maurais (2003) suggests, involves exploring "their relationships and their competition on the world's checkerboard" (p. 13) and adopting a sociolinguistic approach to understanding globalization.

In a global context, the organization of languages and their relationships is elucidated by de Swaan (2001), who emphasizes the role of multilingualism in connecting humanity across diverse languages. He categorizes languages into a hierarchical system: Marginal languages, pivotal languages, and the exceptionally central language, English. Marginal languages, comprising 98% of the world's languages, are spoken by less than 10% of the global population and often lack a writing system. Pivotal languages, like Chinese, Hindi, and Russian, serve as second languages for speakers of marginal languages and hold significance in politics, education, media, and more.

English occupies a central position as the central language, linking supercentral languages together in the global language system. The dominance of English at the core of this system can be attributed to historical facts like colonization, migration, and the leverage of scientific and cultural achievements. Today, English is prevalent in various fields such as literature, science, technology, and international business, holding legal and statutory status in 70 countries (Crystal, 2003). Leitner (1992) further underscores the historical factors contributing to English's central role such as large-scale migration, colonialism, and cultural achievements.

Blommaert (2010) emphasizes that linguistic tools possess economic and social principles that outline their geographical dispersion. Cross-cultural forums and their subsequent linguistic structures are limited by the sociolinguistic aspects related to movement. Languages used in marginal, and intercultural environments are frequently underrated, and using these languages outside of such contexts may be considered problematic.

Balancing Global Competence and Native Language

Children who attain fluency in multiple languages enjoy enduring and substantial intellectual, mental, and societal advantages. These benefits encompass enhanced personal expression, the capacity to empathize with and glean insights from diverse individuals, and improved professional skills as they progress through life.

Within the realm of education, diversity manifests as students entering with life experiences marked by a multitude of languages and cultures, which can be perceived as either a challenge or an advantage (Piccardo et al., 2017). Language is the only mode of socialization as the migrant's languages and cultures differ from those of the host societies. This dynamic creates an opportunity for transformation, with language serving as a potential catalyst. It underscores the importance of emphasizing the connection between multilingualism, cognition, and creativity in educational contexts.

Multilingualism refers to the coexistence of multiple languages within an individual speaker. Plurilingualism, on the other hand, represents the interlinked mastery of numerous linguistic systems and is viewed as a driving force for fostering innovative learning in our modern, highly diverse society. (Piccardo, 2013). Multilingualism is often defined as an individual's ability to communicate proficiently in several languages, achieving a level of fluency comparable to that of a native speaker (Bussmann, 1996). However, in practical written and spoken communication, it's common for proficiency in one language to overshadow the others in a multilingual environment. An alternative viewpoint on multilingualism considers it the simultaneous presence of numerous languages in a given society (Lyons, 1981).

In the context of multilingualism, individuals frequently employ strategies for communication across different languages like code-switching and code-mixing. Code-switching is when individuals use different languages to communicate at home and in public. This is seen mostly in India where people of diverse ethnic and lingual streams live together and we can see the seamless transition between languages such as English, Hindi/Urdu, Bengali, and Tamil (Lyons, 1981). Understanding and conversing in different languages helps people to mingle with the heterogeneous society and also survive in the society. It also provides more employment opportunities to the general public wherein they can showcase their strengths across the nation and the globe.

Language Learning and Identity

The word "language" frequently refers to the mother tongue of an individual which represents his ethnic and lingual background which is different and not the same across the globe. This shows how language is a discrete, self-contained system with defined boundaries. By recognizing one's language we can discern his social identity along with his cultural heritage. Banks (1993, 2002), claims that these identities assign particular characteristics, abilities, and social standing to people.

Additionally, fluency to converse in different languages grooms the intrapersonal, and interpersonal features of a person and also boost his moral, and social conscience, broadening his cognitive domains. It also nurtures respect for diverse cultures from an early age (Burbules & Torres, 2013). This diverse communal interaction becomes an important asset for people to survive in the competitive global world where technology and information transfer are not feasible without language fluency.

Learning a language is an art and it requires a variety of semiotic resources and is a dynamic process. B.F. Skinner (1957) argued that a child acquires language by connecting words with their meanings. According to this perspective, when a child produces language correctly and when a child recognizes that it serves a communicative purpose, this reinforces the language development of children (Ambridge & Lieven, 2011). Chomsky (1965) first suggested the idea of Universal Grammar (UG), which he defined as abstract knowledge of innate grammatical categories (nouns, verbs, etc.) that support language development and processing in general. The innate view states that UG should contain all the grammatical information required for a kid to develop language.

The primary function of education is not only focused on the transaction of knowledge but also enables children to empower their critical thinking and problem-solving skills (McLeod, 2008). This perspective has significant implications for how languages are perceived within classrooms and as subjects of language education. It emphasizes the interconnectedness and fluidity of language acquisition, encouraging a more holistic and integrated methodology for acquiring language skills.

Looking at language through the lens of a national language often leads to the perception that each "language entity" is internally cohesive and, simultaneously, autonomous from other languages, emphasizing their differences. This perspective became more prominent with the rise of nation-states, as the concept of "language" became closely tied to specific territories. Just as geographical borders were established, languages started to be seen as distinct entities separated by boundaries (Hobsbawm, 2004). The discourses of linguistic autonomy and distinctiveness are intricately linked with the ideology of the nation-state and the development of national languages (Joseph, 2006).

Language Policy and Multilingualism

Education should aim to equip individuals with a diverse language repertoire and the knowledge of which languages are most relevant for specific purposes. This underscores the need for launching a strategy for language to enhance the educational program, instructional techniques, and educational encounters within the public educational system within a state. Such a policy should also work toward promoting a more favorable and safeguarded role for multiple languages in society, recognizing their value and importance.

The world has always been inherently multilingual, and our approaches to language learning and teaching must acknowledge this multilingual reality.

Multilingualism has consistently been the natural state of humanity. Across the world, children develop in an environment where they have access to different languages from a young age, and as young individuals pursue their education and careers, they often find themselves in environments where languages other than their first language are prevalent. In such situations, becoming bilingual or multilingual becomes a necessity.

The United Nations (UN) has actively embraced multilingualism, recognizing the importance of using its six recognized languages to enable efficient communication regarding an extensive spectrum of intercontinental matters. Similarly, the European Union (EU) has made multilingualism a fundamental value, with a staggering 24 official languages. This commitment to multilingualism underscores the significance of language in shaping cosmopolitanism as a global, personal, and cultural identity.

Empirical studies have shown that language proficiency indeed plays a crucial role in broadening one's worldview. Importantly, foreign language skills tend to foster greater tolerance and openness to different ideas. A video titled "How Learning a New Language Makes You More Tolerant," showcased by the World Economic Forum in 2017, portrays how language acquisition enhances our comfort when encountering novel experiences and scenarios, thereby enhancing our capability to adeptly handle interactions with novel concepts and different approaches to accomplishing tasks.

The European Union (EU) upholds and supports multilingualism as an integral aspect of its citizens' and nations' cultural heritage. The EU has devised a range of policies to promote and safeguard multilingualism, grounded in mutual or common beliefs concerning the advantages of a heterogeneous linguistic society. The EU's multilingualism policy is structured around three key pillars:

1. **Preservation of Linguistic Diversity**: The EU works towards the preservation of the abundant linguistic variety present in Europe.
2. **Promotion of Language Acquisition and Polyglotism**: The EU actively encourages language learning and the cultivation of multilingualism.
3. **Languages for Trade and Competitiveness**: The EU recognizes the significance of languages in enhancing trade and competitiveness on the global stage.

One of the main goals of the European Union's multilingualism strategy is to ensure that every European citizen can effectively communicate in two languages apart from their native language. The EU's language policy can be succinctly described as follows:

> The goal of EU language policy is to foster the teaching and learning of foreign languages within the EU and to create a welcoming environment for all Member State languages. Proficiency in foreign languages is considered one of the fundamental skills that all EU citizens should

> acquire to enhance their educational and employment prospects within the European learning society. This is especially important in the context of exercising the right to freedom of movement of persons.
>
> *(European Parliament, 2016)*

Language and Globalization

The notion of globalization is a tangible reality because our world has become highly interconnected. This connectivity has been facilitated by advancements in transportation and interaction systems, resulting in increased global interactions regarding trade and culture among nations worldwide. This has led to the complex and frivolous interaction that has become a strong criterion for indulging in the inclusive global community, where communication is the only mode of information transfer verbally and technologically (Sobol, Cleveland & Laroche, 2018). It is only by introducing global languages we can embrace the idea of incorporating a global perspective into the educational curriculum. The vital aim of this system is to build strong connections and positive exchange between students and individuals academically and technically (Goren & Yeminy, 2017).

We inhabit an increasingly complex and globalized world, a reality that has both intriguing and paradoxical consequences for our existence. From one perspective, globalization fosters uniformity by influencing the market and shaping products, tastes, and cultures. Moreover, it gives rise to a broader spectrum of society, with the assertion of local and regional identities and the emergence of societal and cultural disputes.

A noteworthy aspect of globalization is the profound effects of polyglotism and the associated concept of cultural pluralism. In today's world, very few societies can be regarded as culturally or linguistically homogeneous. Diversity prevails, not only in the linguistic landscape but also in the myriad lifestyles and means through which individuals reside and convey themselves, encompassing their cultural expressions

In this modern era, it's crucial to recognize that monolingualism is no longer a strength but rather a limitation (Crystal, 2006). Embracing multilingualism and multiculturalism enables societies to thrive and navigate the complexities of our globalized world more effectively.

Multilingualism is frequently understood as a situation where a population is proficient in one or more national languages and perhaps one or two major languages acquired through formal education. In terms of providing multilingual services, it can often involve offering services primarily in the national language but also including English alternatives, assuming that a significant portion of the population can communicate in English.

In certain African nations, children are proficient in multiple languages even before they commence primary education. They acquire one language within their household, pick up one or more languages within their local community, and subsequently learn a third or even a fourth language as the primary medium of instruction in school. This highlights the rich linguistic diversity and multilingualism that exists in many regions, particularly in Africa, where multiple languages play significant roles in everyday life and education.

Language in the Globalized Economy: Shaping Communication and Competency

Multilingualism is a widespread and increasingly studied phenomenon, attracting considerable scholarly attention in recent times. It is an interdisciplinary concept that can be explored from both individual and societal perspectives. In the 21st century, language skills have become paramount social skills, closely tied to creativity, problem-solving abilities, and effective communication. Understanding teachers' beliefs is crucial for comprehending the way decisions are made within the classroom environment.

In the globally integrated world, the ability to speak foreign languages is seen as a global competence, and being multilingual is recognized as a crucial social aptitude. Multilingualism plays a pivotal role in cultivating the intercultural competencies required for global talent in a rapidly globalizing job market. It also contributes to creating effective and engaged cosmopolitan global citizens and serves as a vital tool in addressing complex global challenges and issues.

There exists a direct and significant connection between the way we communicate in the present day and the dynamics of the globalized economy. The global economy displays three distinctive features (Castells, 2000):

1. **Information-Driven Economic Processes:** It is characterized by economic activities that heavily rely on the generation and utilization of information.
2. **Global Scale of Economic Production:** Economic production has transcended national boundaries, diminished the exclusive dominance of national economies, and operated on a global scale.
3. **Networked Competition:** Competition is structured within networks that span the globe, redefining the traditional notions of competition and market dynamics.

These features, which have evolved and intensified beyond what was previously envisioned even just a decade ago, profoundly influence the way we communicate and, consequently, the role of language. This shift is challenging long-held assumptions about concepts like "community" and "communication."

In this evolving landscape, communication has become both local, often involving multiple languages, and global, characterized by instantaneous and standardized forms of interaction. The interplay between local and global communication is a defining aspect of our contemporary world, reshaping how we connect with others and share information in a rapidly changing globalized society.

Multilingualism and Knowledge Acquisition

In our globally diverse and culturally rich world, proficiency in languages from other cultures and cultural knowledge have become essential global competencies and social skills. They enable individuals to engage with a diverse and interconnected global society more effectively, fostering tolerance, open-mindedness, and the ability to appreciate different perspectives and ideas.

According to Vygotsky, scientific concepts are not isolated entities but are rather part of a broader system of concepts. These concepts are initially unsystematized and infused with personal experiences. It is through the use of language that these concepts become meaningful and gain structure. Vygotsky's perspective underscores the importance of first language usage in teaching and learning within higher education, especially when it comes to transferring knowledge. When students grasp and develop concepts in their native language, it lays the foundation for more effective learning and understanding. This process of concept development can significantly enhance students' chances of achieving academic success as it allows them to construct a deeper understanding of complex ideas and topics.

According to Ladson-Billings (1995), multilingualism is viewed as a scaffolding tool i.e., multilingualism can serve as a valuable scaffold for learners, especially those who are navigating the complexities of multiple languages and cultures.

Promoting Multilingualism: A Societal Imperative

The significance of language extends far beyond the realm of education and educators. This realization has been acknowledged by influential figures in the fields of education and psychology, spanning centuries. Notable among these figures are Comenius from the 16th century and more contemporary thinkers like Jean Piaget, the Swiss psychologist of the 20th century.

Comenius and the Power of Education: In the 16th century, Comenius articulated a profound truth that still holds today: "*A society's fate, in the end, depends on the quality of the schooling its children receive in language.*" Comenius recognized that the educational foundation laid in language profoundly shapes the destiny of a society.

Piaget and the Societal Role of Education: Jean Piaget, a prominent Swiss psychologist, emphasized the integral role of education within the broader

> societal context. According to Piaget, "*Education is not limited to the action of schools and families; it is an integral part of the formative process inherent to all individuals and an essential component of general social life. Human society, in essence, is an educational society.*" Piaget's perspective underscores the idea that education permeates all aspects of social life, playing a central role in influencing the progress of both individuals and society in its entirety.

These insights from Comenius and Piaget highlight that language and education are not isolated facets of society but integral components of the broader social fabric (Comenius, 1947; Piaget, 1993). Recognizing the importance of language and fostering multilingualism is not just a matter for educators; it is a collective societal endeavor that shapes the destiny and progress of nations.

Preparing for a Globalized World

Multinational corporations often select a common language, known as a lingua franca, making it easier for internal conversation among workers in the corporal factions. Several prominent corporations, such as Renault, Samsung, and Airbus, have adopted English for their organizational linguistic approach (Lyall, 2016).

In international sports, multilingualism is the standard practice. Events like the Olympics designate French and English as official languages, in addition to the language of the host country (Lyall, 2016). Many global sports entities tailor their social media content to specific regions, ensuring linguistic relevance. As an example, in 2016, Major League Baseball (MLB) mandated the presence of Spanish translators for all U.S. baseball teams to accommodate Spanish-speaking players and fans (Double Play, 2016; Guardado, 2016).

Multilingualism offers students a valuable tool for deconstructing and reconstructing their learning experiences by drawing upon their various languages as resources. This process not only validates students' existing knowledge and backgrounds but also exposes them to different cultures, enriching their overall understanding. More exactly people don't judge their culture instead they volunteer to explore and recognize the different linguistic cultures all over the world.

This strategy facilitates a proper understanding of the content, ultimately creating an interest in access to the epistemology and history of the discipline. In short, multilingualism encourages students to explore diverse perspectives and develop a more intact knowledge of the discipline, increasing their speed of language acquisition

In the complex and fast-moving technological era, there is a strong need for educational reform which paves the way for incorporating diverse curricula and embracing cross-country perspectives in global education. This transition opens up the significance of breaking down diverse disciplines across the globe and fostering an international perspective. By incorporating globally integrated collaborative projects, students have the opportunity to gain cross-disciplinary knowledge and skills that support them in attaining success in real life. Such projects lead to concrete accomplishments and empower students' thinking and creativity by innovating fresh ideas and concepts that make them competent and also broaden their horizons. They can also improve their thought process and manifest ideologies based on global cultures thus progressing and initiating novel realms toward global prospectus.

Interactive and collaborative learning methods are instrumental in this process. Through this approach, students take on more responsibility for their learning, synthesize knowledge from various subjects, engage in direct communication and activities with individuals who speak different languages, and become adept at working in diverse and multilingual environments. This not only boosts their educational growth but also furnishes them with essential life competencies for the globalized world.

Due to the economic, political, and cultural challenges brought about by globalization, the only passport to engage in the global community while fulfilling the demands of the 21st century is the voluntary embrace of multilingualism and multiculturalism. This inclination, which is increasingly common in developing countries, is also recognized as a necessity in developed nations, including Europe, Asia, Africa, and Latin America. Its rapid spread has been accelerated in response to economic, political, and cultural imperatives. Consequently, the practice of learning languages in addition to one's native tongue is now considered an unavoidable norm.

Conclusion

Proficiency in the utilization of other languages has been associated with individual advantages. These encompass the capacity to relish literature and cinema in their original language, enhancing travel experiences, and enabling communication with individuals globally or locally. Personal benefits extend to emotional intelligence, social adeptness, and the impact of multilingualism on one's cultural identity (Vince, 2016).

Moreover, cognitive advantages encompass an improved understanding of one's primary language and language in a broader context, commonly referred to as metalinguistic awareness. Multilingualism enhances skills such as multitasking, adopting different perspectives, employing analytical cognitive techniques, maintaining cognitive control, making sound financial decisions, nurturing creativity, and encouraging divergent thinking. It has also been correlated with delayed onset of Alzheimer's disease and dementia. In the academic sphere, students in K-12 foreign language programs have demonstrated higher SAT scores.

Understanding multiple languages offers insights into diverse cultural segments, fostering appreciation and comprehension of these cultures. This knowledge is instrumental in the interdisciplinary domain of global understanding and intercultural competence, indicating a well-educated global citizen. It has been associated with emotional, cultural, and cognitive intelligence. It has been recognized as a factor in identifying and solving problems, primarily due to the greater acceptance and understanding of languages. Differences in memory retention among multilingual individuals also contribute to this advantage. However, the most compelling reason lies in the enhanced acceptance and endorsement of solutions.

In essence, within our interconnected and diverse global landscape, multilingualism transcends the boundaries of mere linguistic proficiency; it emerges as a linchpin for unlocking cognitive, cultural, and professional opportunities. It empowers individuals to meaningfully engage with the global community, fostering not only tolerance but also open-mindedness and collaboration across linguistic and cultural frontiers. In closing, this comprehensive exploration reaffirms that multilingualism is not confined to the realm of academic study; rather, it stands as a dynamic and transformative force that is reshaping the very foundations of education, communication, and the essence of global citizenship. Its significance is undeniably profound, and its potential, boundless, as it continues to weave itself into the ever-evolving tapestry of our interconnected world.

References

Ambridge, B., & Lieven, E. V. M. (2011). *Child Language Acquisition: Contrasting Theoretical Approaches*. Cambridge: Cambridge University Press.

Banks, James A. (1993). Multicultural Education. *Phi Delta Kappan*, *75*(1), 21–28.

Banks, James A. (2002). Teaching for Diversity and Unity in a Democratic Multicultural Society. *Education for Democracy: Contexts, Curricula, Assessments*, *2*, 131.

Bairy, S. (2019). Multilingual Approach to Mathematics Education. *Online Submission*, *7*(2), 71–86.

Barnes, D. (1976). *From Communication to Curriculum*. Harmondsworth: Penguin.

Biseth, H. (2009). Multilingualism and Education for Democracy. *International Review of Education*, *55*, 5–20.

Blanc, M. (2008). Multilingualism in the Ancient Near East. Paper presented at Birkbeck College. Unpublished Manuscript.

Block, D. (2005). *Multilingual Identities in a Global City: London stories*. Oxford: Palgrave.

Blommaert, J. (2010). *The Sociolinguistics of Globalization*. Cambridge: Cambridge University Press.

British Council (2011). *Dreams and Realities*. London: British Council.

Bullock Report. 1975. *A Language for Life: Report of the Committee of Inquiry Appointed by the Secretary of State for Education and Science Under the Chairmanship of Sir Alan Bullock*. London: Her majesty's stationary office. [http://www.educationengland.org.uk/documents/bullock/] Date of access: 28 May 2024. Pp.1–437

Burbules, Nicholas C., and Carlos Alberto Torres. (2013). Globalization and Education. New York: Routledge. http://books.google.ie/books?id=FsJdAgAAQBAJ&printsec=frontcover&dq=Burbules,+N.+C.,+%26+Torres,+C.+A.+(2013).+Globalization+and+education:+An+introduction.+In+Globalization+and+education+(pp.+1-26).+Routledge.&hl=&cd=1&source=gbs_api

Bussmann, H. (1996). *Routledge Dictionary of Language and linguistics*. London: Routledge.

Castells, M. (2000). Globalisation, Identity and the State. *Social Dynamics, 26*(1), 5–17.

Chomsky, N. (1965). *Aspects of the Theory of Syntax* (pp. 1–251). Cambridge, MA: MIT Press, 1965.

Comenius, J. (1947). The Development of Modern Education: Comenius. In R.Ulich (Ed.). *Three Thousand Years of Educational Wisdom: Selections from Great Documents* (pp. 339–346). Cambridge, MA: Harvard University Press.

Council of Europe (2007). *From Linguistic Diversity to Plurilingual Education: Guide for the Development of Language Education Policies in Europe*. Strasbourg.

Council of Europe. Council of Europe (2009). Language Education Policy Profiles. Retrieved from: www.coe.int/t/dg4/linguistic/Profils_EN.asp

Council of Europe (n.d.). Retrieved from: www.coe.int/t/dg4/linguistic/Schoollang_EN.asp

Crystal, D. (2003). *English as a Global Language* (3rd Ed.). Cambridge: Cambridge University Press.

Crystal, D. (2006). *How Language Works*. New York: The Overlook Press.

Cummins, J. (2013). Language and Identity in Multilingual Schools: Constructing Evidence-Based Instructional Policies. In D. Little, C. Leung & P. van Avermaet (eds.) *Managing Diversity in Education: Languages, Policies, Pedagogies* (pp. 3–26). Bristol: Blue Ridge Summit: Multilingual Matters. https://doi.org/10.21832/9781783090815-003.

de Swaan, A. (2001). *Words of the World: The Global Language System* (pp. 1–253). Cambridge: Polity Press.

Double Play: Major League Baseball Hires Interpreters, Breaks Language Barriers. (2016). http://www.kjinternational.com/inthe-news/double-play-major-league-baseball-hires-interpreters-breaks-language-barriers (accessed 7/11/16).

Dufva, H., Suni, M., Aro, M., & Salo, O. P. (2011). Languages as Objects of Learning: Language Learning as a Case of Multilingualism. *Apples-Journal of Applied Language Studies, 5*(1), 109–124.

European Commission (2005). Retrieved from: http://eur-lex.europa.eu/legal-content/en/TXT/?uri=CELEX:52005DC0596

European Commission (2008). *A Rewarding Challenge: How Language Diversity Could Strengthen Europe (Proposals from the Group of Intellectuals for Intercultural Dialogue)*. Brussels: European Commission.

European Commission (2012). *First European Survey on Language Competences*. Brussels: European Commission.

European Parliament (2016). Retrieved from: www.europarl.europa.eu/atyourservice/en/displayFtu.html?ftuId=FTU_5.13.6.html

Goren, H., & Yemini, M. (2017). Global Citizenship Education Redefined – A Systematic Review of Empirical Studies on Global Citizenship Education. *International Journal of Educational Research, 82*, 170–183.

Guardado, Maria. (2016). Long Overdue, Spanish-Language Translators a Key Move for MLB. http://www.nj.com/mets/index.ssf/2016/05/mets_spanish-language_translators.html (accessed 7/11/16).

https://www.rajeevelt.com/multilingualism-and-the-power-of-language-national-education-policy-2020/rajeev-ranjan/#:~:text=The%20NEP%202020%20acknowledges%20the,educational%20experience%20by%20advocating%20multilingualism.&text=4.11.

https://www.cambridgescholars.com/resources/pdfs/978-1-4438-4754-4-sample.pdf

Johnson, K. (2001). *An Introduction to Foreign Language Learning and Teaching* (pp. 1–57). Beijing: Foreign Language Teaching and Research Press.

Joseph, J. E. (2006). *Language and Politics*. Edinburgh: Edinburgh University Press. http://cup.columbia.edu/book/978-0-7486-2452-2/language-and-politics

Ladson-Billings, G. (1995). But That's Just Good Teaching! The Case for Culturally Relevant Pedagogy. *Theory into Practice*, *34*(3), 159–165.

Leclerc, J (1999, revised 2016). Retrieved from: www.tlfq.ulaval.ca/AXL/monde/index_politique-lng.htm

Leitner, G. (1992). English as a Pluricentric Language. In M. Clyne (Ed.), *Pluricentric Languages: Differing Norms in Different Nations* (pp. 179–237). Berlin: Mouton de Gruyter.

Little, D. (2010). Languages at School: A Challenge for Multilingual Cities. In King and Carson, 2016.

Lyall, Sarah. (2016). Brazilians Speak Portuguese, but the Olympics Must Use French. http://www.nytimes.com/2016/08/10/sports/olympics/french-official-language-olympicgames.html?action=click&contentCollection=sports&module=NextInCollection&region=Footer&pgtype=article&version=new sevent&rref=collection%2Fnews-event%2Frio-olympics-2016&_r=0 (accessed 7/11/16).

Lyons, J. (1981). Kembo Sure, eds. *African Voices: An Introduction to the Languages and Linguistics of Africa*. Southern Africa: Oxford University Press, 2000.

Lyons, J. (1981). *Language and Linguistics: An Introduction*. Cambridge: Cambridge University Press.

Maurais, J. (2003). Towards a New Linguistic World Order. In J. Maurais & M. A. Morris (Eds.), *Languages in a Globalizing World* (pp.13–36). Cambridge: Cambridge University Press.

McGrew, A. (1992). A Global Society? In S. Hall, D. Held, & A. McGrew (Eds.), *Modernity and Its Futures* (pp. 61–116). Cambridge: Polity Press.

McLeod, A. (2008). *Listening to Children: A Practitioner's Guide* (pp. 1–224). Islington: Jessica Kingsley Publishers.

Ministry of Human Resource Development (2019). National Education Policy, (479 pages). Retrieved from https://mhrd.gov.in/sites/upload_files/mhrd/ files/Draft_NEP_2019_EN_Revised.pdf

Ontario Ministry of Education (1977). Curriculum Guideline for the Senior Division English; Toronto: Ministry of Education.

Piaget, J. (1993). Jan Comenius. In *Prospects* vol. XXIII.

Piccardo, E. (2013). Plurilingualism and Curriculum Design: Toward a Synergic Vision. *Tesol Quarterly*, *47*(3), 600–614. https://doi.org/10.1002/tesq.110

Piccardo, E. (2017). Plurilingualism as a Catalyst for Creativity in Super Diverse Societies: A Systemic Analysis. *Frontiers in Psychology*, *8*, 2169. https://doi.org/10.3389/fpsyg.2017.02169

PROBE (1999). *Public Report on Basic Education*. New Delhi: Oxford University Press. https://www.unesco.org/en/multilingualism-linguistic-diversity

Ridouani, D. (2011). Multilingualism, Multiculturalism and Globalization. *Postmodernism Problems*, *1*(2), 206–222.

Skinner, B. F. (1957). *Verbal Behavior* (pp. 1–33). New York: Appleton-Century-Crofts.

Sobol, K., Cleveland, M., & Laroche, M. (2018). Globalization, National Identity, Biculturalism and Consumer Behaviour: A Longitudinal Study of Dutch Consumers. *Journal of Business Research*, *82*, 340–353.

Stein-Smith, Kathleen. (2016). The Role of Multilingualism in Effectively Addressing Global Issues: The Sustainable Development Goals and Beyond. Theory and Practice in Language Studies. 6. 2254. 10.17507/tpls.0612.03.

Vince, Gaia. (2016). Why Being Bilingual Works Wonders for Your Brain. https://www.theguardian.com/science/2016/aug/07/being-bilingual-good-for-brain-mental-health (accessed 7/11/16).

Wade-Leeuwen, B., Vovers, J., & Silk, M. (2018). Explainer: What's the Difference Between STEM and STEAM? *The Conversation.*

Yakman, G. (2017). STEAM Education Professional Development Practicum & Research. *Proceedings of the PATT, 34.*

Chapter 20

E-Learning – A Way for Bridging the Digital Divide in Andaman and Nicobar Islands

Opportunities and Challenges

K. Venkatesan

Introduction

E-learning is not new. However, the COVID outbreak has propelled schools and educational institutions to drive e-learning initiatives aggressively. Traditional classrooms have their limitations, and innovative digital platforms for learning are shaping our education systems in a big way. With crisis come challenges. It's time to realize the full potential of e-learning.

The CBSE and Ministry of HRD directed the Department of Education, Andaman & Nicobar Administration to take several e-learning initiatives and mitigate students' fears during and beyond the pandemic. As a result, the Directorate of Education provided students in classes II to XII with learning content through the local Doordarshan channel, All India Radio, and cable network.

Apart from this, many e-learning platform development companies sprouted in the islands to provide state-of-the-art learning solutions to businesses. They provide web-based training through Learning Management Systems (LMS), mobile apps, and exam portals. Another common approach includes learning theory from home

DOI: 10.1201/9781003504894-20

and attending practicals at training centers. This combination of online and offline saves students and educators time.

We'll learn about the various initiatives the Government and educators took to bridge the digital divide in the Andaman and Nicobar Islands.

The Background

Inequality in education does not exist anywhere in the remote areas of Andaman and Nicobar Islands. The stakeholders in the islands including Scheduled Tribes have done their enrollment to obtain formal education through schools. Their enrollment rate is high. To take them from traditional to digital hubs of learning E-learning is a must but requires the bare minimum of facilities to operate. While on one side, there is adequate infrastructure for the remote area learners, there's a vast digital divide on the other end. Opportunities are lost due to a lack of proper internet facilities in the islands. However, the administration and students are both enthusiastic about e-learning. Online education empowers learners to showcase their skills in an open, interactive, and advanced environment. In addition, certified online courses from reputed institutions can help generate quality employment opportunities.

What Is E-Learning?

E-learning involves any learning activity with the help of digital resources such as a computer and the internet. It transfers skill and knowledge over an internet connection. Education is delivered to many students in real time or at different times.

When initially introduced, e-learning took its share of beating and was not wholeheartedly accepted. Both students and parents complained about the missing human element. But, over time, with rapid enhancement in technology and learning systems, the masses gradually embraced it.

Embracing the Digital Age

With a background of demographic shock and cultural collapse, the local population of the Andamans is always apprehensive of any global change. Lack of respect and resources have already created a massive education gap. Cultural discrimination and schools lacking basic infrastructure are often significant obstacles to equality in education. In addition, school violence, gender discrimination, and sometimes sexual abuse contribute to high dropout rates.

To make matters worse, the pandemic, lockdown, and social distancing have pushed the traditional education system into a tight spot. So, in a nutshell, the pandemic-hit businesses and a weak educational structure are somewhat forced to embrace the digital revolution. We often have no choice but to ride the wave. The Indian education system had to approve and implement pre-recorded sessions and live-streaming classes. Besides, challenges and trials are always part of any profound change, and educators are currently navigating potential solutions.

How Can E-Learning Revolutionize Education in the Chain of Islands?

Electronic learning empowers students to learn without attending a school with better curricula and organization. It can reach the remotest areas and is accessible to all. In addition, experts believe that the audio-visual form of education engages students. E-learning ensures limitless advantages to students because of technological advancements.

Students in the Andamans look for alternative education methods to mitigate budget cuts, high tuition fees, and course shortages. E-learning courses are continually improving, and their certifications are of equal value to on-campus ones. As a result, the number of people enrolling in online programs is rising daily.

Let's explore how digital learning is impacting the lives of students.

E-Learning Considers the Needs of Every Student

Online education caters to everyone's needs. Whether it's books or newspapers, modern technology has transformed how we consume content. The digital sphere has become all-inclusive. Earlier, people with regular jobs, along with family responsibilities, couldn't think of upgrading their skills and knowledge. And the same goes for homemakers. But, thanks to the digital revolution. Now everyone can take online classes during their productive hours, be it a housewife or someone in a daily job.

Busy lifestyles and intense daily schedules have made multitasking the new norm. So, whenever people have some time to spare, maybe a weekend or otherwise, online programs are there to help them upgrade.

Lower Cost

Cost-effectiveness is one of the key benefits of e-learning. Educational institutions are saving loads of money on accommodation and transportation for students and educators. Besides, students don't have to buy pricey textbooks, which means even more savings. Moreover, the entire process is eco-friendly as there is no need to print anything.

Remote Approach

Not so long ago, the only way to pursue an academic course was to go to a college or university physically. With all the top colleges and courses in Port Blair, students from far-flung towns or islands had to travel daily or relocate. Today, besides the traditional three or four-year programs, colleges offer various online courses that students never had access to before.

In the current scenario, students don't have to visit campus and adhere to protocols and fixed schedules. Instead, education is brought to them. Any course or program in their field of interest is accessible if they are willing. In addition, students can access the study material as often as they want.

With the online approach, students become masters of their own time. They are free to pursue any education related to their interest, which was unthinkable under the traditional curriculum. Not to mention, students don't have to commute in harsh weather during dangerous travel conditions. Instead, they can attend class, participate in chat sessions, join discussion boards, and turn in their homework on time without missing out on lectures or deadlines.

Rapid Career Growth

E-learning keeps students in sync with the latest updated content from promising lecturers worldwide. Moreover, online courses give a ray of hope to anyone who had to discontinue their academic studies for various reasons and then never got a chance to pursue their higher education after that.

Online courses encourage professionals to further advance in their current jobs or apply for new ones. In addition, an e-learning certificate can prove their professional dedication and prowess to prospective employers.

Faster Delivery

Online learning provides faster lesson delivery when compared to a traditional classroom structure. In addition, it reduces learning time by 30% to 60%.

a. Online Lessons are direct and go straight into the subject. Most topics are wrapped up within the same learning session enabling the e-learning program to roll out within a few weeks or even days.
b. Self-paced programs allow students to progress at their own speed.
c. Students save on commute time.
d. Learners have the choice to skip lessons that they're already accustomed to without wasting time.

Consistent and Scalable

Traditional classroom lectures were vulnerable to mistakes, as they heavily depended on a lecturer's temperament. E-learning eliminates such loopholes due to the standardization during a course-creation process. Moreover, consistency is maintained to ensure equal distribution of knowledge among all candidates. Likewise, the local shortage of experienced teachers will never be a problem again because a handful of highly qualified educators can teach thousands of students online.

E-learning can positively influence every business in the islands. Employees who have taken e-courses are better equipped to face any upcoming challenges as they're well in tune with the latest and the greatest. Their efficiency will reduce business expenses. The workforce can enhance their abilities and implement their gained knowledge into daily tasks.

Personalization and Analytics

Each student has a distinct learning style. And that's why the traditional one-size-fits-all teaching approach is not the favorite among students. A unified solution can't match everyone's expectations. Instead, E-learning offers various learning styles for students to adapt to their desired classes. No matter what their background is, there's a class for everyone. Moreover, students can pursue higher education without attending any unwanted courses. Online learning caters to each learner's preferences and goals.

E-learning platforms constantly collect student data and analyze them to improve study materials and boost training outcomes. Various processes monitor student activities and dropout rates to identify potential shortcomings in the course materials. They can further track whether or not the implemented changes have brought about any improvement. Data collected from a learner's activity is invaluable, and its full potential is yet to be realized.

Key Challenges to Overcome

The geographical profile of this vast archipelago makes it dependent on the mainland for all essential commodities. Resources are always tight due to a lack of sizeable freshwater bodies. Education in the islands is still a struggle as the attitude of the Indian government is colonial at its core. Social hierarchy is strong enough to decide both administrative attention and resource allocation. Furthermore, the distant islands are far out of reach and at local bodies' mercy. The problem with education in the Andamans is unique because the society there is highly decentralized.

Geographical Challenge: Port Blair is the only town among the islands that is better off regarding educational infrastructure and opportunities, but the distant places are less fortunate. Education is an expensive commodity in the far-flung islands. Furthermore, the tsunami that struck the Indian subcontinent in December 2004 compounded the problems of access, retention, quality, and infrastructure.

a. **Financial bottleneck**: The inhabitants of the islands are not alien to the benefits of education. Even the tribal communities of the Nicobar Islands send their children to state-run schools and are aware of its importance. However, most of the population are daily laborers with at least three children per family. So, educating everyone with limited resources severely strains the family's economy. Although primary schooling is free, additional expenses for procuring uniforms, stationeries, etc., are high.
b. **Network irregularity**: Although 4G is available on the islands, it seldom works. Uploading a single file could take hours, let alone attending live-streaming classes. While Port Blair has intermittent access to the internet, the speed elsewhere is abysmally low and practically nonexistent in the Nicobar Islands. As a result, far-flung islanders may see online learning processes as cumbersome and less user-friendly. Inconsistent Zoom or WhatsApp calls can easily distract and irritate a student.
c. **Device Cost:** The island's IT infrastructure is not as robust as the mainland. Moreover, all sections of people there are not capable enough to procure all the necessary devices to aid online learning. Both educators and learners require devices such as modems, routers, smartphones, or laptops. In addition, special software and tools are sometimes required to purchase along with the course, which can add to the cost.
d. **Lack of engagement**: The physical absence of the teacher often impacts the learner's performance. Students sometimes are not motivated to engage with lessons remotely and cannot continue their coursework. Also, instructors or supervisors find it difficult to monitor learning progress.
e. **New learning environment**: For many people, the idea of wrapping their heads around learning in personal spaces or at home is still new. Traditionally, schools, colleges, or training centers have been considered learning venues, whereas living rooms or bedrooms are for rest or leisure. So, when surrounded by household chores, family, and pets, the learner's mindset is affected. This lack of focus makes the initial struggle even more challenging.
f. **Conflicts with responsibilities and time**: Life is tough on the islands. Children after a certain age often have to work alongside other adults in the family to supplement their income. So, not every student has the luxury to incorporate learning into their schedule when they're already working.

Opportunities that E-learning Offers

Commutation is a massive problem in the islands. Most habitation is along the island's coastal strips, and outside of Port Blair, the public transport system is weak. As a result, students often have to walk for six to seven kilometers if they miss a bus. Moreover, private-run jeeps are expensive and don't follow a set schedule. So, in a scenario where getting to school is a significant challenge, e-learning is a saving grace.

Certain pockets of the population don't send their children to school to avoid heavy dents on the family income. Moreover, it's not only the kind of education available but also the quality that matters. There is often only one college on the island that offers basic sciences and arts. On top of it, a lack of opportunities coupled with pressing financial needs forces students to drop out. However, online learning is lighter on the pocket and provides the students with the desired skills at their doorstep.

A lack of digital curriculum and internet connectivity is yet another problem in the far-flung islands. Low-quality digital education directly impacts a learner's prospects. Online courses can mitigate all the risks we discussed. In-house specialists design E-learning programs. Shorter videos divided into shorter segments ensure a longer attention span for students. Limiting each class to fewer students enhances engagement and teacher presence. Learners no longer have to travel overseas for courses as there are alternatives available on reputed online platforms.

Types of E-Learning

Factors like learning method, delivery style, and educational tools are used to decide the types of e-learning.

a. **Computer-Managed Learning (CML)**: CML systems use computers to teach and assess learner performance. The process includes developing tests, analyzing them, and keeping a progress report. Finally, the ranking algorithm adjusts the learning process based on student preference. Additionally, CML stores training materials, lecture information, and curriculum information.
b. **Computer-Assisted Instruction ((CAI)**: CAI uses the best of both traditional teaching and computer-based learning. In addition to conventional teacher-directed instructions, CAI includes simulation activities, tutorials, and drill-and-practice. CAI methods are preferred for skill development programs because interactivity is high, and students remain more active during the process.
c. **Synchronous Online Learning**: This method employs videoconferencing and live chat to teach a group of students simultaneously in real time. Instructors and students interact without any delay. Synchronous learning is the fastest-growing e-learning method because it eliminates social isolation and boosts teacher-student relationships.
d. **Asynchronous Online Learning**: In contrast, asynchronous learning takes a different path. It allows learners to study independently in different locations and time zones without real-time interaction. In addition, students have more flexibility with this approach. Asynchronous teaching methods include eBooks, blogs, discussion forums, emails, CDs, and DVDs.

e. **Adaptive Learning**: As the name implies, this teaching method constantly adapts and redesigns learning materials to fit the needs of each learner. The system creates more student-centered tools by considering individual parameters such as goals, abilities, and performance. In addition, technology can be used to assess higher-level skills more effectively using laboratory-based techniques. Moreover, artificial intelligence (AI) plays a primary role in shaping adaptive learning to personalize the learning experience. The main modules of an adaptive learning system are knowledge management and retrieval.

f. **Linear eLearning**: This method allows a one-way communication where the information or data is passed from the sender to the receiver. The sender sends the information after determining the time, order, and speed. There is no feedback process.

g. **Interactive eLearning**: Unlike linear e-learning, this system allows two-way communication. The sender and receiver are interchangeable. Discussion boards, forums, and instant messaging are some examples. The interactive approach enables learners and instructors to modify learning and teaching methods.

h. **Individual eLearning**: In this approach, students study the material independently and are expected to meet learning objectives. The method mirrors the traditional classroom system. Finally, peer grading or automated evaluation evaluates learners' performance against learning objectives.

i. **Collaborative eLearning**: On the contrary, in collaborative e-learning, two or more students form a group and engage in the learning process. This learning methodology aligns with the modern approach that best comes out in a group setting. Individuals interact and learn from each other. They operate about each other's strengths and weaknesses.

How Is the Indian Government Helping the Andamans?

Remote Learning Initiatives

The Directorate of Education took many initiatives to promote learning from home among school students.

Local Doordarshan Broadcast

The Department of Education and Doordarshan, Port Blair, broadcasted pre-recorded teleclasses for higher-secondary students throughout the islands in 2020. The key highlights of the classes are as follows:

a. A one-hour slot was assigned on all weekday evenings except weekends.
b. The study material covered only Class X and XII subjects.

c. Class X subjects included English, Science, Maths, and Social Science.
d. Class XII subjects included English, Accountancy, Geography, Physics, Chemistry, Mathematics, Economics, Political Science, Business Studies, Computer Science, Biology, and History.
e. The classes followed the NCERT syllabus.
f. Classes used PowerPoint presentations and other digital tools.
g. Doordarshan telecasted the program across the length and breadth of the Andaman and Nicobar Islands.

Tele-Classes on Local Cable TV Network

The Department of Education planned to broadcast pre-recorded online classes through local cable TV channels. This particular scheme was aimed at students of classes VIII to XII.

a. It was initially planned as a one-hour slot at noon but was increased later to a three-hour slot.
b. HD recordings of the same contents were repeated telecasted in the evening.
c. Good quality equipment and professional videographers were involved.
d. Teachers used PowerPoint presentations and different digital tools.
e. South Andaman District was the main target of the telecast. However, the cable TV network repeated the classes' telecast to other places as well.
f. The program covered every crucial subject for each class.
g. Educators provided the students with phone numbers and email IDs to handle queries regarding the topics.

Radio Classes

As alternative learning, All India Radio, Port Blair, started audio classes for classes II to VII.

Initially, the program held sessions for classes V to VII. Later, the broadcast also added lessons for classes II to IV.

a. A one-hour slot was assigned for all six days except Sundays.
b. Teachers covered major subjects such as English, Hindi, Social, Science, and EVS.
c. Educators took the necessary steps to create easy-to-digest materials for students in primary classes.
d. The education program kept Saturdays for doubt-clearing sessions.
e. They also introduced the "Kitabon Se Parey Ki Rochak Duniya" feature on Saturdays to cover additional subjects like Moral Education, Scientific facts, Computer knowledge, GK, etc.

Providing Digital Learning Contents Online

Online Videos on YouTube Channel

A YouTube channel called "AN Education Dept Andaman" was set up, and all the recorded teleclasses were uploaded.

a. Whoever missed the classes got an opportunity to see them online later.
b. The educators uploaded quite a few videos on different subjects.
c. The channel received a good response with hundreds of views.
d. The teachers conducted Tele-poster and Tele-quiz competitions.
e. Students and educators formed WhatsApp groups to interact.

Digital Content Repository

Looking at the growing demand for study materials from students and parents alike, the Department set up a Common Service Centres (CSCs) repository. Digital content is stored in the State-owned Data Centres and publicly available through CSCs. The Department plans to develop an e-content Digital Library in collaboration with the State Institute of Education.

Samagra Shiksha

The Department of Education has been trying to provide quality education and an intelligent classroom environment to the students of the A&N islands. As part of the initiative, the department planned to set up smart classrooms in 162 government schools across the islands under the Samagra Shiksha scheme. Despite Covid challenges, the department successfully implemented smart classrooms in 141 government schools. In addition, the authorities trained the teachers to get acquainted with the devices and make learning more attractive, commutative, and easy to grasp. Samagra Shiksha fostered opportunities for educators and learners by integrating learning technologies like computers, software, assistive learning devices, audience response technology, networking, and audio/visual platforms. Smart classrooms attracted students and proved beneficial in remote areas to enhance learning with advanced technology.

The Andamans Get High-Speed Net

In August 2022, The Prime Minister of India inaugurated the country's first undersea optical fiber cable network between Chennai and the Andamans. The 2,300-kilometer submarine fiber cable is 5G ready. The connection promises 200 Gbps connectivity between Port Blair and other islands. The undersea cable is expected to usher in an education revolution on the islands and offer the people a reasonable shot at e-learning initiatives. The long-awaited project will create a sense of confidence and integration among the educators and learners on the islands.

Moreover, improved digital connectivity raises hopes regarding newer learning possibilities and job opportunities.

Future of E-Learning

The islands face many socioeconomic challenges, and online learning has to take a different approach. WhatsApp is more successful than other technological platforms because of its simplicity, low cost, efficiency, accessibility, and instantaneous access by students and adults alike. So, educational authorities must keep this in mind while designing e-learning platforms in the future. The success of any online learning program depends on the following.

a. Technology is not the only requirement for effective remote learning.
b. Teachers play a more crucial role than ever. Therefore, developing a healthy teacher-student relationship is vital.
c. Meaningful two-way interaction between students and teachers will allow growth to happen.
d. Parents play an equally important role in mitigating the limitations of e-learning. Authorities must guide parents and equip them with the necessary tools to help them support students.
e. The Department of Education needs to collaborate with other multi-lateral, academic, private, and public entities to orchestrate various players and promote a quality learning experience.

Conclusion

Some islanders don't have the education level and financial status to teach their children at home or provide internet facilities. The government and local bodies have to come together and ensure that nothing closes the door to quality education. They must spread awareness among students and adults and encourage everyone to embrace the digital revolution. Authorities must strive to make learning seamless and immune to poverty or pandemic. Increasing internet bandwidth, improving capacity-building measures, including intranet, and using electronic devices effectively are a few steps to empower e-learning among students of the Andaman and Nicobar Islands. Every system has flaws, but we cannot deny the greater good.

References

https://leadschool.in/blog/growing-importance-of-e-learning-in-the-21st-century/#:~:text=Online%20learning%20helps%20students%20to,and%20teachers%20develop%20advanced%20skills

http://14.139.60.153/bitstream/123456789/4430/1/Programme%20of%20action%20for%20education%20for%20all%20in%20andaman%20and%20nicobar%20islands%20D8841.pdf

https://vikaspedia.in/education/education-best-practices/remote-learning-initiatives-in-india/remote-learning-initiatives-in-andaman-nicobar-islands

https://talentedge.com/blog/scope-opportunities-online-learning-india/

https://elearningindustry.com/how-india-fare-on-global-elearning-spectrum

https://ailit.org/top-10-elearning-benefits/

http://www.legalservicesindia.com/article/2177/Fundamental-Right-of-The-Child-To-Education-in--Andaman-And-Nicobar-Islands.html

https://www.livemint.com/news/india/what-internet-means-to-an-island-economy-11598278668106.html

https://www.edapp.com/blog/remote-learning-challenges/

https://dt.andaman.gov.in/DetailNews.aspx?newsid=OCw7Si9Hzdcv+u+fCY7dWC4l5zPQNSMy/GpM7H1CCKA=

https://research.com/education/what-is-elearning

Chapter 21

Digital Financial Transaction Practices among Urban and Rural Women

R. Pavitra and Kamalaveni

Introduction

Digital literacy is now an important aspect of every individual across the globe, be it a student, professional or homemaker. Digital literacy is not an option but a life skill that improves efficiency at work, access to services, communication, collaboration, expression and advocacy. In the present scenario, Digital literacy plays a crucial role in the empowerment of women, by easing tasks, providing updated information on world issues, encouraging networking and seeking support. Owing to the ease and comfort of completing the otherwise tedious tasks from the comfort of the home, almost all sectors like banking, education, health and marketing, resorted to access customers digitally. Women who were aware of and adapted to the use of technology managed to adjust to the changes while others had to depend on their children, partners or friends to sustain their life with ease. GSMA found that 393 million adult women in developing countries do not own mobile phones and globally women are 8% less likely to own a mobile phone than men. (https://www.unicef.org). It has become mandatory that skill education has to go hand in hand with digitalization. Children and elders alike need to be educated to use the various applications required for daily tasks like payment portals (Paytm, Google

DOI: 10.1201/9781003504894-21

Pay), meeting rooms (zoom), ordering items online, etc. In the current study, four women in Coimbatore District interviewed to understand the skills they acquired to use technology.

Digital literacy involves the understanding and utilization of digital tools and technologies for performing meaningful operations like bank transactions and purchase online things daily. For example: the use of a personal computer, laptop, smartphones and other devices encompasses digital literacy. It is also referred to as an individual's ability to find, evaluate and communicate information through typing and other media on various digital platforms (Sarpong, 2021). Digital literacy today is an essential life skill that is almost a necessity for survival in the technological era. Digital skills are one of the new demands for the people in the current generation which cannot exclude any gender, caste, age or sex.

It is a skill that mandates individuals to change, learn, and adapt to a new atmosphere. People need these skills to maneuver through various applications and programs available to not only be active on social media but also complete necessary tasks like shopping, banking, payments and transactions. Digitalization of information has transformed labour processes and markets from a physical to a virtual mode while telecommunications have enabled relocations of jobs globally. The increase of information technology across all economic activities has led to an increase in e-skills not only for IT professionals but also for other workers in other sectors. With this prelude, the current study aims to examine the digital skill practices among urban and rural women for handling financial transactions.

Objective

To examine the digital skill practices among urban and rural women for handling financial transactions

Statement of the Problem

With greater access women have started to reap the benefits of the technology revolution as there are various initiatives in the recent days that offer services to help women increase their connectivity and to help them teach how to utilize the internet (Tapashi, November, 2018). India has the second largest online market, with over 140 million internet users. Despite the great potential, women are heavily underrepresented in the digital world, particularly in rural settings, where access, cost and other socio-cultural factors limit the abilities of women. The research study by UNICEF shows that there is a large gap (33% of women use mobile phones against 67% of men) between the male and female users of mobile devices. Due to this, India's rank has fallen to 21st place in the Global Gender Equality Report published by the World Economic Forum and ranked 108th out of 144 nations. This seems to represent a new challenge for gender equality as digital skills are unequally distributed and very smaller number of educated women are more as advantaged as men.

Benefits of improved and increased access are gained by only a segment of our society that is equipped with digital literacy, and this disadvantage becomes more striking while considering the gender divide as well as the urban-rural divide. Due to the gender roles and associated issues, problems of labour in India affect women and men differently. Difficulties in accessing technology resulting from inadequate technical means are faced much more by rural communities as digital resources are unevenly distributed between the urban and rural areas. Nationwide surveys have shown that India has a lack of digital literacy which is leaving a large segment of Indians behind as the rest of the country goes online.

The NFHS survey showed that only 42.6% of the Indian women surveyed had ever used the internet, compared to 62.16% of the men. According to the GSMA report, 2021 that analyzed the gender gap in access to mobile phone, 79% of the adult male population and 67% adult female population are mobile phone owners in India (Shahnahan, 2022).

Technological and Innovative Advancements in India during Covid-19

There have been many technological and innovative advancements in the country in the past couple of decades. The Digital India program was launched by Prime Minister Narendra Modi on July 1, 2015, following which Digital advancements have reached the remotest corners of the country. The e-governance program and advantages of the same have reached the lives of people of all ages, religions, castes and sexes in the country. The advent of digitalization has been a blessing to most of the population. In these years, people from the remotest corners have started functioning on digital platforms, especially for financial transactions like Paytm, Paypal, GooglePay, and other apps which can be easily handled on any smartphone. In rural India, the availability of digital resources does not ensure access to the internet. In most villages, accessing the internet daily, at a high cost is considered more of a luxury than a need. Especially during emergencies like the lockdown of the pandemic, families chose to buy items of daily use for the families rather than pay for the internet to access classes for the children or provide internet facilities for the women. Increasing internet use expands opportunities for Hybrid and remote work, financial transactions, education and video calls, but the pervasive gender gaps inhibit women's ability to realize these opportunities (OECD, 2018).

When the Pandemic struck India, it affected the lives of every person in the country. After the initial shock and confusion, people found ways to adapt to the situation. The onset of the pandemic made digital technology an important facet and a necessary competence that is a requirement of today's life. Being tech-savvy or Digi-smart was a big advantage. All those who were able to use digital devices to their advantage were able to bridge the gap created by the pandemic, much better. Students were able to attend classes online and organizations offered the work-from-home or remote working option. In this context women in rural and urban areas were mandated to learn certain digital skills for survival.

Status of Rural and Urban Women in Post Covid Period

The technological advancement during the pandemic was accompanied by certain pressures. The masculine domination in the use of finance and resources as well as established norms in the rural social milieu where women do not try to be smart or display their knowledge base as against their partner, are also important reasons why many women either do not receive smartphones, are scared to request for one and even if one is accessible are not confident to use the same due to fear of rejection, ridicule or anger (Sarpong, 2021).

In some rural India families, women do not own smartphones either due to lack of finance or due to familial constraints. Other women were able to access the same but did not want to use due to lack of training, knowledge, awareness of the same and they stagnated for some time in their life. Many South Indian women in rural areas still use the initial feature phone or base models of mobile phones with non-touch displays. These phones support only limited actions like voice calls and SMS messages.

On the other hand, many women who could afford a smartphone and whose functioning needed a mandatory use of technology were able to purchase a smartphone and become more tech-smart. They were able to learn to access many applications that were required for completing immediate tasks like online payments and banking. They took the effort to learn the use of technology. It is seen in the towns as well as the villages, that the largest supermarkets, the street vendors, the autos and cabs, or even the largest establishments in post COVID, function with UPI payment or wallet payment schemes like g-pay or Paytm. These are user-friendly, easy to use cash to digital payments, do not involve physical cash transactions and are secure. Another advantage is smallest transactions amount. Most vendors and small businessmen turned to these payment gateways and only very few vendors prefer cash transactions. According to Kalpana Ajayan in developing India from the *Times of India*, women face challenges due to responsibilities in private sphere and restricted access to information and technical skills, lack access to smart phones and digital financial services. The fear of making a mistake is so deeply construed in the women that they would much rather yield their independence and allow the men to make decisions and transactions rather than take the risk of making a financial mistake is another barrier for women. In evolution of digital era, it is important to understand how widely women are open to learning digital transaction skills, how considerable support do they get from family members to equip themselves with digital transaction skills and the benefits they get by acquiring the skills. In this context the current research aims to examine the digital skill practices among urban and rural women for handling financial transactions.

Methodology

Method of Study: Qualitative approach using in-depth interview

Data collection Tool: dimensions focusing on accessibility to digital resources, awareness on Digital financial applications, type of Digital Financial Applications used (like banking app, shopping apps, educational apps, UPI apps, etc, situation of usage of digital apps, digital learning strategies employed by the women, level of Support (from family, education, society) and struggles involved were studied. The current study uses both primary and secondary data. The study discusses the interviews held with four women hailing from rural areas in Coimbatore district. They were selected using snow ball technique for the study. In-depth interview method was used to analyze the usage of digital technology pattern among women. The study is qualitative in nature. Comparative analysis between rural and urban women is done in the current study.

Rajitha

The first respondent Rajitha was 43-year-old lady, employed as full-time home maid in an urban city. She was educated up to 10th std, married, and had 2 college-going children, who stayed with her in-laws. Her husband was working as a fisherman, collecting fish from the wholesale market and selling it on his two-wheeler. He lost his job during the pandemic when free movement and door-to-door trade were restricted. He was an alcohol addict and had no savings for the family. Rajitha was a homemaker and with so many dependents and limited resources, Rajitha resorted first to work in a grocery store near her residence. The store was small in a rural area and the income was limited. It was not enough to pay for her children's education, in-laws medicines and family maintenance. She decided to get new job and signed up for an agency providing manpower/home nurses to various homes. Through the agency she got placed in a home for a 24-hour duty as a care taker. Alcoholic nature of her husband hard-pressed Rajitha to open a bank account and handle her finances by her own. She also got a smartphone to take care of her payments and other needs of the family. She linked her bank account and within two months she learnt to use payment portals like g-pay and Paytm. She learned to complete online transactions for her daughters' fees and college needs. She ordered online things to meet her family needs during her absence. She also learnt to order medicines online and other necessities for the family. Rajitha says that she has gained digital knowledge with the help of her owner lady where she is employed. The independent decision making on financial matters was not appreciated by her husband where he sees it as a form of insubordination. However, she was supported by her in-laws. She says "I am thankful to the Covid that forced me into a situation of independence otherwise would have continued to be dependent on my husband who treated the family so badly." The present interview reveals that even though Rajitha did not own a smartphone and wasn't financially independent, she was able acquire the digital skill when there was need. Though she was fearful to learn the financial transactions in the beginning, she learned it over the time. Her minimal education helped her to read the basic instructions on the phone and her workplace support helped her to overcome the situation. Examining the digital literacy skills in financial transactions of Rajitha shows that she was more open to learning the digital financial transaction and got support from extended families to learn the skill. Her learning supports the children's education and enhanced her independence. She was not aware of various apps like UPI, Paytm and GPay in the beginning but later slowly she equipped herself with this knowledge. Also, her move from rural area to urban area exposed her to the digital practices around her. That gave her confidence to learn easily. She was smart enough to acquire the basic digital transaction skills.

Bhavitha

The second respondent Bhavitha was married, 24-year-old unemployed homemaker, with 10th grade education. She had a seven-year-old son with her husband and mother-in-law in an urban town. (Lost the father-in-law recently to COVID). She was using smartphone for the last 6 months. Her husband was employed as a goldsmith and went for other odd jobs. She used the phone to access WhatsApp to get the important information shared from her son's school. Though Bhavitha was educated till her 10th grade, she was not very confident about using the smartphone. She got smart phone when her husband got a new one for himself. She has an Instagram account which she doesn't understand many things. She revealed that her husband was ready to support her in assisting how to use Paytm and G-pay. Her husband wants her to be independent in small transactions. But Bhavitha was reluctant to learn due to the fear of losing it. Her husband constantly encourages her informing how these gateways can help in cashless journey. Bhavitha was so rigid in learning digital financial transaction skills which was noticeable through her statement that "I will hold financial transactions with friends and family or banking, on an offline basis rather than online" when I need it immediately. She also prefers to do online shopping via POD (pay-on-delivery) basis, to avoid possible fraudsters. Though she seems to be knowledgeable about media and also has a supportive husband, Bhavitha is suspicious of handling financial transactions online. She is not sure that any of the transactions are safe and can be trusted. She used the smartphone for entertainment and watches YouTube to learn new recipes and also uploaded a few of her videos on the same. Examination of the digital transaction skill shows that Bhavitha is reluctant in learning the new skill because of the fear of losing money however the usage of social media like what's app and you tube for entertainment shows the expansion of the digital world.

Bhavya

Bhavya was married 22-year-old unemployed lady with 2nd p.u.c grade education, has a three-year-old son and one-year-old daughter. She stayed with her husband and in-laws in a semi-rural village on the outskirts of the city. She got smartphone for past three years when her husband got employed as an auto driver and earned moderately well. She knew how to use the smartphone make calls, do WhatsApp, has media interaction on Facebook and Instagram and watches YouTube videos. But she doesn't use Paytm or G-pay or make any other financial transactions. Online shopping was done from her husband's phone so that payment can be taken care of by him. Bhavya can read English and can follow instructions and she was active on social media and checks her phone often. Her husband is a supportive person and has opened a bank account for her. Bhavya prefers to make only cash payments and is cautious of online payment portals. On further questioning she says that she prefers her husband to handle all the finances of the family because she is feared of being confronted if something goes wrong. She has never handled large amount of cash even before marriage and so she is scared of making financial transactions. She is not confident about handling financial transaction via digital mode. The present case study shows that the educational status alone cannot decide the financial transaction practices but it the is inherent interest and willingness to learn it. Though she has learnt to handle the smartphone she did not want to take risks. The current case study also shows that women need to be familiarized with the decision-making process on family matters right from childhood and free of preconceived notions that only men handle finance well. The gender stereotyping of women's role has withheld her to learn new skill involving risk.

Chandrika

Chandrika was a 48-year-old married woman, hailed from a rural village and presently staying on her own in a semi-urban town. Her husband was an alcoholic and had an affair with another woman due to which she was separated from him two years back but yet to be divorced. Her three daughters are married and having children of their own and are settled with their respective in-laws. Chandrika studied till 6th std many years ago and had to discontinue studies due to financial constraints in the family.

Chandrika was earlier working as a housemaid but lost her job during the pandemic. The household where she was working was initially worried about infection but later, they lost their jobs and hence couldn't afford a home maid. In a bid to support herself and her family, Chandrika took up selling vegetables. She took permission from the panchayat and started selling vegetables that she bought from the wholesale market, near the bus stand. She was sure that people who returned from work would see the vegetables and buy them. Her idea was successful and her shop did good business. She found that holding cash in small change was difficult and had security issues with transporting money at late hours. Her estranged husband had once tried to take cash from her when she was returning home after selling vegetables. Hence, she got a smartphone, downloaded the necessary applications and learned from her neighbor to make and receive payments on Paytm and g-pay. This has been of immense help to her. She says now she can't go wrong in calculations and she has learned to use the calculator on the phone. When speaking with her, she said that she used the earlier phone for making calls, while she used the smartphone specifically for her business. She said she is hopeful of being able to learn how to operate the rest of the applications available on the phone, to use them to her advantage in her business. Even though Chandrika is only educated till primary school, she has been able to learn digital transaction skill. Even though she may not have learned to operate or understand the various applications on the smartphone, she has been able to learn digital skills to ease her work and has found efficient ways to function. She has been able to adapt well due to the training she has received. The interest she has in learning, advancing her business, and improving her financial affairs has been a driving force that has helped her to learn digital skills that have enabled her to be independent.

Discussion

Evolution of technology in last two decades has necessitated each individual to learn digital skills for the day-to-day survival. The ability of a person to be able to access and navigate the digital space will determine how well one can adjust, and able to sustain the information that is relevant in the ever-changing world. It is to be noted from the case studies of the women mentioned in the current study, that digital skills and technical literacy have become an absolute necessity in one's life.

India is the second largest populous country in the world after China and has a majority of its increasing population in its working age group. This gives India the added advantage of surpassing China by 2025 in terms of its economic development as a more working population implies more development (Shahnahan, 2022). But this demands that the 50% of its population be equally productive, and on a smart basis at par with the rest of the world.

Technology can help women overcome hurdles with timely access to the right and relevant information, market, mentoring, capital and customers. Penetration of mobile phones in rural areas could be used to bring these women closer to the stakeholders. India has over 63 million micro, small and medium enterprises, half of which are in rural areas and approximately 20% of these rural enterprises are owned by women. Till now, they are accustomed to dealing solely in cash. Making these businesses digital would help women merchants cope better with the current crisis as well as similar upcoming ones. However, there is a need for these women to be first provided with connectivity, including mobile phones, and taught digital skills professionally.

Proper training in using digital space is required so that women are not only taught to handle the various applications they need but also the correct ways of handling these. Correct knowledge of the pros and cons of using digital space, and awareness of the various issues that can come up like hacking into accounts, scams of asking for details via phone calls, etc are necessary so that women do not get cheated and remain wary. Losses can be avoided and confidence can be built with more and correct information. When training is being provided, especially in the rural sectors, it is to be noted that the language needs to be specific. Not all users may be aware of English. Therefore, women should be taught how to change the modes and choice of language on their system (Laptop, mobile, etc) to their language of comfort. This makes it easier for women to learn and identify with the new tasks on hand. In the place of teaching/learning, the person teaching should be trustworthy, safe, and approachable to check with in case of doubts. It is also necessary to keep in mind that some of the women may be disabled. Hence, teaching should be inclusive and adaptive.

Recommendations

The study showed that of the four women, the women who were educated and had more chances of using digital apps (respondents 2 and 3) were the people who used it lesser than the two women who were less educated. This could be mainly due to circumstances demanding the change for survival as well as the support of the immediate family members. Respondents one and four were in situations that demanded independence and they received the support of family to learn the use of the device and apply the knowledge to everyday living.

The current research shows that despite the age and lesser education level, with proper support the two women were able to learn and apply their knowledge. Understanding from the same one can expect similar development and positive changes in the lives of women in rural India.

The change that is expected with a positive outcome forecasted, is possible only if the factors of change are multi-pronged.

- The individual's effort to learn and adapt is of foremost importance. If women don't take the onus of learning upon themselves, the change stops.
- Digital skill education can be parallel with formal school education.
- It must be made mandatory to undertake digital skill training for the women willing to take up entrepreneurship before lending the loan for their business activities.

- The family is the primary support provider to all individuals. When a woman is trying to achieve something different, the family needs to make an effort or make adjustments to allow her the comfort of a learning mode. It is evident that only with the influence of male -breadwinners of the family that women take up to learn new skills. It needs to be relooked at in a way that unpaid labour skills should be shared and technical skills should be learned by women.
- The strength of the individual is possible to be undermined, or effort to be diluted if societal perspectives do not change its patriarchal notions. Women need the support of the society they live in to learn, change and grow. Education entails not only academic or scholastic gain but also skills that need to be developed by people to empower them, regardless of age or gender and women need to be supported to be digitally smart.
- NGOs are known to have brought about changes that even governments thought were difficult to make. More NGOs need to work on the provisions/ease of access to technology in the rural areas as well as involve in training rural women to use the financial digital skills.
- Government can proposal special schemes to enhance financial transactions digital skills among rural women

Conclusion

Examination of the digital skill practices among urban and rural women for handling financial transactions is imperative need in the evolving the digital era. Understanding women's role in handling digital transactions will help the policy makers to design the programs needed to empower rural women in handling financial transactions via digital space. The research study concludes that irrespective of educational status among women, it is significant to motivate women to use cashless transactions. The digital transactions should be familiarized among women to reduce the fear of losing money. The scope of digital space should move from social media usage to life skill usage.

References

OECD. (2018). *Bridging the Digital Gender Divide.* Paris: OECD Publishing.

Sarpong, E. (2021). Beyond the Basics: Quality, Speed, Affordability, Relevance. In OECD, *Development Co-operation Report: Shaping a Just Digital Transformation.* Paris: OECD Publishing.

Shalmahan, M. (2022). *The Mobile Gender Gap Report.* London: GSMA.

Tapashi, D. (November, 2018, November). A Study on Aspects of India's Digital Divide. *Research Review International Journal of Multidisciplinary*, 3(11), 689–693. http://www.rrjournals.com

Websites

https://www.westernsydney.edu.au/studysmart/home/study_skills_guides/digital_literacy/what_is_digital_literacy

https://www.shethepeople.tv/news/impact-technology-powerful-women-rural-india/

https://www.nationalheraldindia.com/india/so-much-for-digital-india-only-426-of-women-6216-men-have-ever-used-internet-finds-govt-survey

https://www.hindustantimes.com/ht-insight/gender-equality/the-digital-divide-and-is-it-holding-back-women-in-india-101641971745195.html

https://www.unicef.org/eap/media/8311/file/What%20we%20know%20about%20the%20gender%20digital%20divide%20for%20girls:%20A%20literature%20review.pdf

https://news.climate.columbia.edu/2023/03/07/a-just-transition-for-women-working-toward-digital-literacy-in-india/

https://www.oecd-ilibrary.org/sites/bc7b62dc-en/index.html?itemId=/content/component/bc7b62dc-en#back-endnotea27z6

https://thewire.in/economy/digital-india-women-technology

https://timesofindia.indiatimes.com/blogs/developing-contemporary-india/making-digital-financial-services-work-for-women/

Index

Note: **Bold** page numbers refer to tables and *italic* page numbers refer to figures.

Made in the USA
Las Vegas, NV
19 September 2026

6415d758-a906-4121-93e9-444fdce87497R01